EveryDay Geography
of the World

EveryDay Geography of the World

An Entertaining Review of the Land, Climate, People & History of Our World

Kevin McKinney

Illustrations by Michael Moran

BLACK DOG
& LEVENTHAL
PUBLISHERS
NEW YORK

Copyright © 1993 by Byron Preiss Visual Publications, Inc.

Library of Congress Cataloging-in-Publication Data on file at Black Dog & Leventhal Publishers, Inc. Available upon request.

Published by
Black Dog & Leventhal Publishers
151 West 19th Street
New York, NY 10011

Distributed by
Workman Publishing Company
708 Broadway
New York, NY 10003

Text by Kevin McKinney
Illustrations by Michael Moran
Jacket design by Sheila Hart

Manufactured in the United States of America.

ISBN: 1-57912-326-0

g f e d c b a

Contents

List of Maps

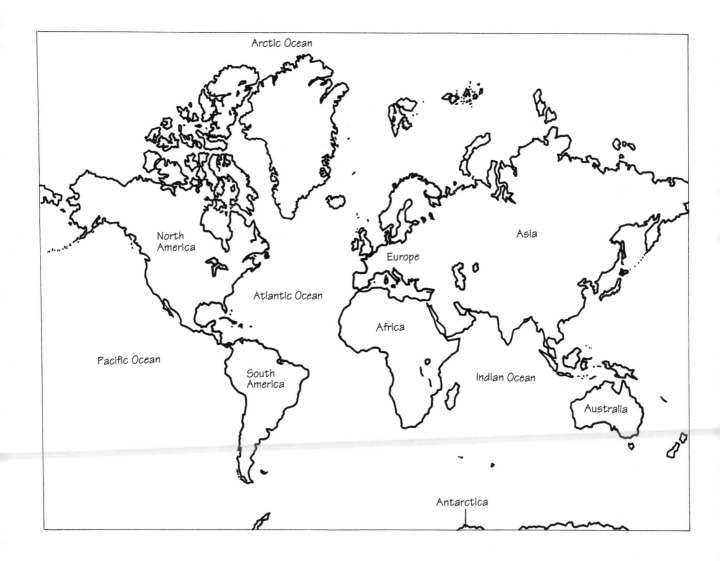

FINDING YOUR PLACE IN THE WORLD

If you were to look down on Earth from outer space, as an astronaut might, the planet would resemble a watery blue gem. Viewed from the blackness of space, the orb is not divided by international boundaries into individual countries, but appears as a cohesive, almost lyrical entity. Indeed, on the planet itself, everything is interrelated. Together, the atmosphere (the gases that surround the planet, including those that make up the air we breathe), the hydrosphere (including the bodies of water that cover the earth), and the lithosphere (the outer hard surface of the planet, generally considered to be about 50 miles thick) make life possible on the earth. The word *geography*, derived from the Greek words *geo* ("earth") and *graphein* ("to write or carve"), refers to the scientific study of Earth—its land, water, and air, and the distribution of plant, animal, and human life. Geographers study such interrelationships as climate and vegetation, atmosphere and the sun, water and soil, and human industries and the

earth. Geography affects every part of your life and helps you to understand the interrelationship between everyone and everything. Agriculture relies on arable land and growing conditions for the food we eat; industry depends on raw materials for use in manufacturing. Understanding geography gives you a better perspective of events on the evening news and provides you with a greater appreciation of your own country's history. Even your reading comprehension will be increased if you have, for example, a basic understanding of the Mississippi River when you read Mark Twain's *Tom Sawyer* or some knowl-

Devastating Changes

• The *Titanic* sank on its maiden voyage after ramming an iceberg in the North Atlantic.

• Deforestation in tropical rain forests causes land erosion and kills valuable medicinal plants as well as humans.

• The *Exxon Valdez* ran aground on a sandbar off the coast of Alaska and damaged the ship's hull, spilling millions of gallons of crude oil that caused the death of wildlife.

• An earthquake in California in 1994 caused millions of dollars in property damage.

• Lava inundated everything in its path when the dormant volcano Mt. Pinatubo in the Philippines suddenly erupted.

edge of Russia before you read Boris Pasternak's novel *Doctor Zhivago*. Moreover, it's important to know not only where you're going and how to get there when you plan a vacation or business trip, but also something about the local culture, language, weather, and political climate of the country. This book provides some quick, practical, even entertaining, answers and insights about geography and our world.

Earth is a dynamic planet that has been going through physical, biological, and chemical changes since its beginning hundreds of millions of years ago. Some changes occur in a matter of decades, others over immense spans of time. Mountains—like those along the west coast of the United States, which rose during the last one or two million years—continue to rise today, the result of the earth's dynamics beneath the surface. Volcanoes spew molten rock and lava from the planet's core, redesigning the landscape or creating islands like Surtsey, Earth's newest island, off the coast of Iceland, and the island of Hawaii, which is still developing. For millions of years, landmasses have been inundated by oceans that recede and leave behind remnants, like Utah's Great Salt Lake. Glaciers carved lakes and valleys in the eastern United States and the Alpine region of Europe, while the glacier covering Greenland has only recently begun to recede.

But you don't need a glacier or a volcano in your backyard to observe geographical changes. Just pay attention to the weather and the seasons. Or watch bulldozers change the geography of an area they're clearing for a 12-lane highway or a 50-story commercial building.

While we can observe the gradual erosion of hillsides and beaches over the course of many years, you and I can't really see activity beneath the earth's surface. But if we're in the right place at the right time, we can experience the effects in the form of earthquakes and volcanic eruptions.

Continents and Landmasses

Rising above the seas, Earth's landmasses take the form of continents—North America, South America, Asia, Europe, Africa, Australia, and Antarctica—and islands. Islands range in size from 2,175,600 square kilometers, or 840,000 square miles, for Greenland, which is the largest island in the world, to less than 1 acre. All of them are completely surrounded by water. Tahiti, Martinique, Iceland, and many other islands are the tops of volcanoes pushed up from the ocean floor. Other islands are actually parts of continents, separated from the mainlands by water but physically connected underwater. These so-called continental islands include Great Britain, which is part of Europe, as well as Newfoundland and Greenland, which are considered to be part of North America.

The earth's crust is broken, like a jigsaw puzzle, into tectonic plates that make up the lithosphere, the roughly 50-mile-thick rock that forms the earth's crust. They move on a hot, viscous mantle, some grinding past each other. Others crash into each other, often forcing one of the plates to slide beneath the other, causing mountains to rise and generating earthquakes.

The plate tectonics theory is a refinement of the continental drift theory, proposed as early as the sixteenth century by Francis Bacon but refined by the German scientist Alfred Wegener in the early 1900s. According to this theory, some 200 million years ago, the earth's landmasses formed a single supercontinent. Dubbed Pangaea by scientists, it broke into two continents, one in the Northern Hemisphere and the other in the Southern. The rifting continued and eventually resulted in the formation of today's continents, which were thought to be more or less floating on the seas.

Part of the support for the continental drift theory lay in the physical appearance of the continents: The contours of the eastern coast of South America and the western coast of Africa, for example, appeared to fit neatly together, like pieces of a jigsaw puzzle. Today we know that the continents are actually the exposed parts of the tectonic plates and it is the plates that move, taking the continents with them, if only inches per century, with the continents growing closer or farther apart, the oceans becoming wider or narrower, and mountains continuing to rise as plates collide. Moving toward the Eurasian plate, for example, the Indian-Australian plate, which carries Australia and India, rammed India into Asia and forced up the Himalayas from undersea rock.

In most cases, the shape of a continent has been determined by the location of relatively young mountains like the Himalayas, which possess the world's highest peaks and are still rising. North and South America also have high, more recently created mountain ranges—the Rocky Mountains and the Andes—in the west and older worn-down mountains in the east—the Appalachians and the Brazilian Highlands—with wide plains between them.

Fortunately, we don't feel the movement of the continents, any more than we feel Earth rotating on its axis or revolving around the sun. Yet the plates continue to bump and grind against each other and, yes, there will eventually be more break-ups. Some doomsayers argue that, in 100 or more years, California will separate from the rest of the contiguous United States, splitting apart at the San Andreas fault, which runs from the Gulf of California to the San Francisco peninsula in northern California.

While we generally refer to seven continents, there are technically only six. Europe and Asia really form a single continent on the Eurasian plate, with Europe being a peninsula of Asia. Only history justifies the

boundary between Europe and Asia, formed by the Ural Mountains, the Ural River, the Caspian Sea, and the Caucasus Mountains. As a result, most of the Commonwealth of Independent States lies east of this artificial boundary, although the most populated areas are on the western (European) side. This makes the Commonwealth the only country in the world that is actually located on two continents. (Asia is also separated from Africa at the Sinai

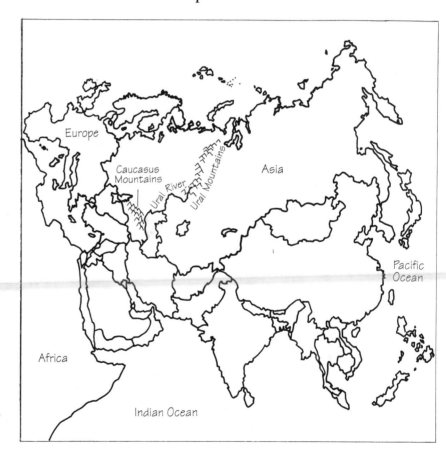

4

Peninsula, where a common landscape extends east and west of a boundary that humans, not nature, determined.)

Geographically, it might also seem more appropriate to consider North America and South America as one continent, since they are connected by a land bridge (Central America). Hundreds of thousands of years ago, the land bridge was flooded, resulting in the separation of the Americas. When the waters receded and land was exposed, ancient peoples were able to cross from North America to South America and populate the southern continent.

While geographers mark the actual separation with the political boundary between Panama and Colombia (where the isthmus of Panama ends), North America is usually considered as the landmass from the Arctic to the southern border of Mexico; South America extends from the isthmus of Panama to the southern tip of Chile, leaving Central America as an anomaly. Much of Central America, however, rests on its own tectonic plate (the Caribbean plate), like Arabia and India.

Despite their geographical boundaries, North America and South America are often viewed in terms of common language, namely English in Anglo America (the United States and Canada) and Spanish in Latin America (Mexico, Central America, and most of South America). The Caribbean islands are also often included with Latin America, although Spanish is spoken on only three major islands—Cuba, Puerto Rico, and half of Hispaniola. Like the residents of the small non-Spanish-speaking Guyana, Suriname, and French Guiana in northeastern South America, people on most of the Caribbean islands speak English, French, or Dutch.

Oceans and Seas

Even if you add up all of Earth's landmasses, they make up only a third of the planet's total surface (about 196,950,000 square miles). The other two-thirds are covered by water—primarily oceans, rivers, lakes, and seas.

Through evaporation, the oceans provide most of the water that the atmosphere eventually returns as rain and other precipitation. Their drifts and currents affect the weather in other ways as well.

Ancient people generally did not venture far beyond the coastlines of their lands. For them, the prospect of oceangoing was frightening because they believed nothing lay beyond the horizon that marked the edge of the ocean and, therefore, Earth. For the most part, and contrary to popular misconceptions, such beliefs had disappeared by the fifteenth century—when European explorers navigated and charted the continent of Africa, and men like Christopher Columbus set sail across the Atlantic Ocean,

The Continental Shelf

On most maps, shorelines distinguish between a continent's landmass and the ocean. Parts of each continent, however, remain underwater. A featureless plain dusted with sands, silt, and mud, the continental shelf extends for an average 72 kilometers (45 miles) from the low-water line. The variation of a continental shelf's width can be drastic: The shelf off the east coast of the United States, for example, is 75 miles, while off the west coast it's only 20 miles. The shelf slopes gradually toward the sea until it abruptly breaks to deep sea. This break is the top of the continental slope that drops to great ocean depths and resembles the side of a mountain. The continental rise lies between the bottom of the slope and the actual seafloor.

Geofact

The world's longest underwater mountain range is the Mid-Atlantic Ridge, extending from Iceland almost to Antarctica, down the center of the Atlantic Ocean.

seeking a western (and shorter) passage to Japan and other lands in the Far East.

Huge bodies of salt water dotted with islands and confined by the continents, Earth's oceans include the Atlantic, Pacific, Indian, and Arctic. Although many people refer to the Antarctic, or Southern, Ocean as Earth's fifth ocean, it is actually the southern extensions of the Pacific, Atlantic, and Indian oceans.

The Pacific is the world's largest ocean, larger than all the landmasses combined. It extends from the west coasts of North America and South America to the east coasts of Australia and Asia, encompassing an area of more than 64,000,000 square miles. The Pacific Ocean is often divided into the North Pacific, from the equator to the Arctic Circle, and the South Pacific, from the equator to Antarctica. The deepest point is the more than 36,000-foot-deep Challenger Deep in the Mariana Trench, west of the Mariana Islands, an independent group of 14 islands east of the Philippines.

South of Argentina, the Drake Passage separates the Pacific from the Atlantic Ocean, which is bounded by the continents of North America and South America on the west and Africa and Europe on the east, covering an area of more than 33,000,000 square miles. Like the Pacific, the Atlantic is divided by the equator—the South Atlantic, from the equator to Antarctica, and the North Atlantic, from the equator to the Arctic Circle. The ocean's deepest point is the Puerto Rico Trench (nearly 23,400 feet), north of the island of Puerto Rico.

Arbitrarily separated from the Atlantic at 20° east longitude, south of Africa, is the Indian Ocean. Spanning more than 28,300,000 square miles, it is bounded by eastern Africa, southern Asia, western Australia and Indonesia, and northern Antarctica. Its deepest point is the Java Trench (more than 25,300 feet deep), south of Java.

North of the Arctic Circle, at about 70° north latitude, the 5,400,000-square-mile Arctic Ocean is almost entirely enclosed by North America, Europe, and Asia. Its greatest known depth, about 18,000 feet, lies in the North Polar Basin, north of Greenland. Parts of the Arctic Ocean are singled out as seas—the Beaufort Sea, Lincoln Sea, Greenland Sea, Norwegian Sea, Barents Sea, Kara Sea, Laptev Sea, East Siberian Sea, and Chukchi Sea—although they don't precisely fit the definition of a sea.

A sea is a segment of an ocean enclosed by land on three sides. Two seas played important roles in the history of Europe. Ancient Phoenicians, Greeks, and other seafarers traveled the Mediterranean Sea (bounded by Europe, Asia, and Africa), establishing Carthage and other colonies in North Africa and along the southern coast of Europe. In the Western Hemisphere, the Caribbean Sea (bounded by North America, South America, and Central America) was the gateway to the New World, containing more but generally smaller islands than the Mediterranean. Many of the earliest European colonies lay in the Caribbean and later served as stopping posts to and from the Americas.

Although it accounts for two-thirds of the planet's surface, water is distributed unevenly around the world—covering 64 percent of the Northern Hemisphere and 74 percent of the Southern Hemisphere.

Other Bodies of Water

A river is a large stream of water that empties into an ocean, a sea, a lake, or another river. Rivers range from the Nile in

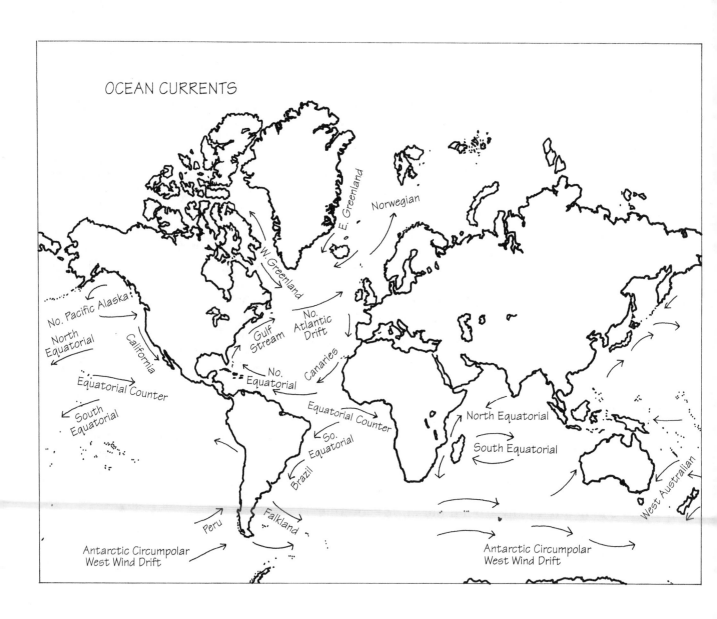

OCEAN CURRENTS

Norwegian

E. Greenland

W. Greenland

No. Pacific Alaska

North Equatorial

California

No. Atlantic Drift

Gulf Stream

Canaries

Equatorial Counter

South Equatorial

No. Equatorial

Equatorial Counter

So. Equatorial

Brazil

North Equatorial

South Equatorial

West Australian

Peru

Falkland

Antarctic Circumpolar
West Wind Drift

Antarctic Circumpolar
West Wind Drift

Africa, which is nearly 4,200 miles and the world's longest river, to the 120-foot D River in Oregon, the shortest. Unlike the salt water in the oceans, river water supplies drinking water for people around the world. In some countries, a river's waters have been declared sacred, as with the Ganges River in India. Many of the large rivers in Russia and Canada flow northward to the Arctic Ocean and, with the long winter season, have little use as a means of transportation because they're always frozen. In such areas, ancient people learned to cut holes in the ice to fish for their food. Rivers are found on every continent except Antarctica.

Lakes are also found everywhere except Antarctica and often provide drinking water. Since they are inland bodies of water generally formed by obstructed water flow, lakes do not always contain fresh water. Examples include the Caspian Sea, a landlocked body of salt water, and Utah's Great Salt Lake, both remnants of the seas that once covered their entire regions. Other bodies of water include straits, harbors, gulfs, and bays.

A narrow waterway, a strait (or channel) connects two large bodies of water. For example, the Bering Strait, west of Alaska, connects the Bering Sea and the Arctic Ocean. The English Channel (in France, called La Manche), which lies between England and France, connects the Atlantic Ocean and the North Sea.

Harbors are inlets or branches of seas, lakes, and other bodies of water where ships can anchor.

A gulf is an area of a sea that reaches inland, like the Gulf of Mexico in the Caribbean Sea, or the Gulf of Oman and the Persian Gulf, northwest of the Arabian Sea.

A wide inlet smaller than a gulf is called a bay. It is usually a part of a sea or lake indenting the shoreline. A bay can be a flooded river valley. In Maryland, on the east coast of the United States, the Susquehanna River once flowed directly to the Atlantic, but today its water fills the Chesapeake Bay.

How to Read a Map

The ability to read a map is one of the most important practical skills a person can have—and you deal with maps more often than you'd think.

Road maps, office building floor plans, and other maps help you determine where you're

River Facts

Even though the Nile is the world's longest river (4,180 miles), South America's Amazon carries the greatest volume of water on its way from its source in the Andes, across Brazil, to the Atlantic Ocean near the equator, a distance of roughly 4,000 miles. In fact, the Amazon contains nearly one-fifth of all the fresh water that runs over the surface of the earth—more water than the Nile, Yangtze, and Mississippi rivers combined.

Lake Facts

The world's lowest: the Dead Sea (nearly 1,300 feet below sea level), a saltwater lake on the Israel-Jordan border.

The world's highest: the Andes's Lake Titicaca, at 12,507 feet above sea level, on the Bolivia-Peru border.

The world's largest saltwater lake: the Caspian Sea, straddling Europe and Asia, a 143,550-square-mile saltwater lake 86 feet below sea level.

The world's largest freshwater lake: North America's 31,800-square-mile Lake Superior.

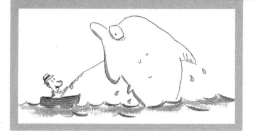

going and the best possible route for getting there. A map can provide the exact dimensions and borders of your property. There are also maps of the political borders of countries or states, maps detailing the routes of public transportation systems, maps of zip codes and telephone area codes, maps of population densities, maps of the planets and the stars, and maps of physical characteristics (topography) of a region, to name a few. You can find maps of individual cities, states, or countries—their borders, topography, population, and other details—collected in an atlas (a book of maps).

It seems that people have always had a need to know where they are in relation to the rest of the world, to determine the physical boundaries between themselves and their neighbors, to know the extent of their territory. Early African tribesmen and Arctic Eskimos, for example, made drawings in the earth or on stone or animal skins to indicate the locations of landmarks so that they or others might find certain places again. Ancient Egyptians and Babylonians drew maps to set boundaries of landholdings in order to collect taxes efficiently. During the Middle Ages, European mapmakers began expanding their maps beyond the continent of Europe, and in the thirteenth century, they began creating world maps, usually based only on what they believed to lay beyond their own familiar territory.

During the last 500 years, mapmakers, or cartographers, have mapped almost every inch of the earth, including the ocean floor. And today, satellite imagery and computers aid the accuracy of maps. Map software and geographic data bases are increasingly available for personal computers, and they make it easier to update or correct maps. (For example, maps sometimes contain geographical errors; political borders also change, as you'll find out in later chapters.) Software and databases enable you to use maps to study geography in relation to other subjects like ecology or history rather than using a different map for each subject.

Unlike a globe, which represents the spherical shape of the planet, maps are two-dimensional graphic representations—from simple street maps to complex navigational charts. They usually include symbols to represent such facts as varying topography and the location of rivers and other bodies of water. Vegetation and climate are usually differentiated by color: Land is usually depicted as green, mountains as brown, water as blue.

No map can offer an actual representation of the territory covered. A map's scale provides the relationship between the length measured on the map and the corresponding distance on the earth's surface. Actual distances may be, for example, 50 times longer, or areas 50 times larger, than they appear on a map. One inch may be equal to, say, 100 miles. To determine the actual distance or area, you would have to determine the number of inches and multiply by 100. The scale is often expressed as a fraction, called the "representative fraction," and is usually written as 1:100 or 1/100, meaning one unit on the map represents (but isn't necessarily equal to) 100 units of distance on Earth. If 1 inch equals 1,000 miles on a particular map and Los Angeles and New York, for example, are 3,000 miles apart, the distance between the two cities on the map would be only 3 inches.

Distances on a two-dimensional map, however, often vary, if only slightly, because the earth is round. Moreover, a map may represent the distance between two points as, say, 1,000 miles, but that's 1,000 miles as the birds fly; it doesn't take

Antipodal Points

In a variation on digging a hole to China, visualize the path of your antipodal point—the straight line from where you are standing directly through the earth to the opposite point on the other side of the globe. If you were able to dig through to the other side, how often do you think you would strike land and how often would you land in water?

With two-thirds of the world covered by water, you might guess that you'd strike water at least 66 percent of the time. It's likely, however, that you would strike water every time. For example, digging from anywhere in Australia, you would emerge in the Atlantic Ocean, where the entire continent neatly fits.

Only a few square miles of the United States have antipodal points on land: Southeastern Colorado corresponds to St. Paul and Amsterdam islands in the Indian Ocean, and a spot on the Montana-Canada border corresponds to Kerguélen Island in the Indian Ocean. If you began digging in the

right place at Point Barrow, Alaska, you'd emerge in Antarctica.

All the Hawaiian Islands, however, are opposite Botswana, Africa. And in South America, Bogotá, Quito, Lima, and much of Chile and Argentina lie opposite China, Sumatra, and neighboring lands.

World's Largest Map

In early 1991, Philadelphia's World Game Society completed the world's largest, most accurate map, measuring 70 feet by 35 feet. In relation to the map's scale, the average man would be about 2,000 miles tall, with his foot about 300 miles long. The moon would be the height of a 70-

story building above the map, and the sun would be about 47 miles away. Prepared from jet navigational charts done by the U.S. Defense Department's Defense Mapping Agency, the Big Map also shows areas of global warming and other environmental problems.

into account the distance of traveling up and down or around mountains.

Latitude and Longitude

Longitude is always measured east or west of the prime meridian. Equivalent to the equator, the prime meridian is an imaginary line that runs north and south through the Royal Astronomical Observatory in Greenwich, England, and divides the planet into the Eastern and Western hemispheres.

Longitude and latitude (sometimes referred to as meridian and parallel) are both measured in degrees, minutes, and seconds: 360 degrees make a full circle; 60 minutes equal 1 degree, just as 60 minutes make 1 hour, and 60 seconds make 1 minute. Degrees are represented by the symbol °, minutes by ', and seconds by ".

Longitude, imaginary lines that encircle the globe from north to south, and latitude, running east and west, are the principal means of pinpointing location on a map. Latitude is always measured north and south of the equator. An imaginary line around the circumference of the planet, the equator is equidistant at all points from the North and South poles and divides the earth into the Northern and Southern hemispheres. The Northern Hemisphere runs from the equator to the North Pole, at the northernmost point of Earth in the Arctic Circle; the Southern Hemisphere runs from the equator to the South Pole, the planet's southernmost point on the continent of Antarctica.

In terms of cartography (mapmaking), you would find Beijing, Madrid, and Philadelphia at roughly 40° north latitude. More precisely, Philadelphia is 39° 57′ north latitude, 75° 10′ west longitude; Madrid, 40° 26′ north latitude, 3° 42′

west longitude; and Beijing, 39° 55′ north latitude, 116° 25′ east longitude.

Because Beijing, Madrid, and Philadelphia are located at roughly the same latitude, all three cities generally share similar climates, as do Cairo, Egypt, and New Orleans, Louisiana, which are both at approximately 30° north latitude. Cairo is at 30° 2′ north latitude, 31° 21′ east longitude; New Orleans, 29° 57′ north latitude, 90° 4′ west longitude.

It depends on your point of view

Most people view the world from their own individual perspective. It's often helpful, however, to look at things through other people's eyes. We were all taught that the earth rotates counterclockwise on its axis and that it orbits the sun taking a counterclockwise path. This, however, only seems to be the case when viewed from the North Pole, at the "top of the world." Earth, in fact, is a sphere spinning in space and has no top or bottom. We think of the North Pole as the top of the world simply because early mapmakers drew it that way, putting themselves on top of the world and everyone else beneath them. If the early mapmakers had been Australian, the continent we refer to as "down under" could easily have been "on top of the world."

Quiz (answers, p. 17)

Before you proceed on the geographical tour of the world in the following chapters, test your knowledge of geography with this quiz. There's no grading system here or for other quizzes in the book. The questions and answers will help you develop an appreciation of the broad range of topics that pertain to geography and encourage you to learn more. You can use the maps throughout this book to answer some of the questions.

1. On what continent would you find each of the following?
   a. the Andes
   b. Notre Dame cathedral
   c. Mt. Everest
   d. the Sphinx
   e. Mt. McKinley
   f. the South Pole
   g. the Outback

2. In what body of water will you find each of the following?
   a. Hawaii
   b. Singapore
   c. Haiti
   d. the Great Barrier Reef
   e. the Bermuda Triangle
   f. Sicily
   g. the Channel Islands

3. On what continent (or island) are you most likely to find each of these endangered species?
   a. giant panda
   b. lemur
   c. grizzly bear
   d. giant tortoise
   e. Bengal tiger

4. Yucatán and Malay are examples of what geographic feature?

5. What modern country was the principal part of ancient Mesopotamia and the "cradle of civilization"?

6. The equator crosses what continent(s)?

7. Most of Africa is located in what hemisphere?

8. If you fly from Rio de Janeiro to New York City, in which cardinal, or main, direction do you travel—west or north?

9. You're going skiing in the Alps. Name at least one country you could be visiting.

10. Where would you find the greatest population of reindeer?

11. Tropical rain forests have the world's largest diversity of plant and animal species. What tropical rain forest comprises the largest land area and where is it located?

12. On what continent would you find the town of Timbuktu?

13. What is considered the world's oldest inhabited city and oldest capital in the world?

14. In Charles Dickens's *A Tale of Two Cities*, what body of water did the characters have to cross to get from Paris to London?

15. The river that inspired Johann Strauss's "Blue Danube" waltz flows through which countries?

16. In 1867 the United States government purchased Alaska from

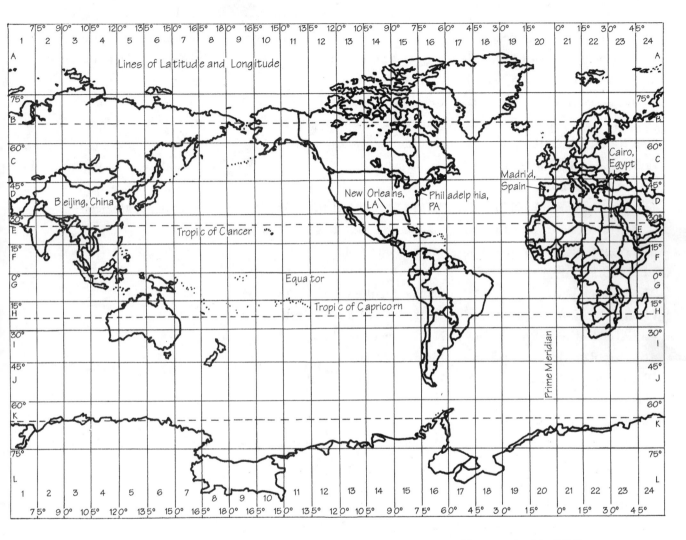

Lines of Latitude and Longitude

Russia for $7.2 million, at a price of two cents per acre. The territory was admitted to the Union as the forty-ninth state in 1959.

a. What body of water separates Alaska from Russia?

b. What country separates Alaska from the contiguous United States?

c. The name Alaska has a geographical origin. What is it?

17. Great Britain's lease on Hong Kong expired in 1997.

a. On what continent is Hong Kong?

b. What is that continent's largest country, which reacquired Hong Kong?

15

18. After it was erected in 1961, the Berlin Wall quickly became a symbol of what Winston Churchill called the Iron Curtain, which divided eastern and western Europe. Twenty-nine years later, when it was dismantled, the wall symbolized the disintegration of the Communist bloc.

    a. From the end of World War II until the reunification of Germany in 1990, where was Berlin located—West Germany or East Germany?

    b. In 1989, if you had addressed a letter to the German Democratic Republic, would you have been mailing it to West Germany or East Germany?

    c. What is the official name of reunified Germany?

19. Lithuania, Latvia, and Estonia are collectively known by what name, based on their geographical location?

20. From 1867 to 1918 much of eastern Europe comprised the Austro-Hungarian empire. At the end of World War I, the dual monarchy was politically divided, with most of its territory becoming parts of other countries. Most of these countries became part of the Communist bloc after World War II. Besides Hungary, name at least one of the Communist bloc countries.

21. If you wanted to attend the Sorbonne, what language would you need to speak?

22. Piloting the *Spirit of St. Louis* on a 1923 nonstop solo flight from North America to Europe, Charles Lindbergh changed the world. What body of water did he have to cross?

23. What is the geographical origin of the numerical system we use today?

24. How many of the five Great Lakes border the state of Michigan?

25. Using the world map on page 15, determine the nearest longitude and latitude for the city where you live. Try to find some other city at the same latitude in your hemisphere, and another on the same longitude.

    Now identify each of the seven continents and the five oceans. Then locate Greenland, Japan, New Zealand, Madagascar, the Mediterranean Sea, the North Sea, and the Gulf of Mexico.

ANSWERS

1. a. South America
   b. Europe
   c. Asia
   d. Africa
   e. North America
   f. Antarctica
   g. Australia
2. a. Pacific Ocean
   b. South China Sea
   c. Caribbean Sea
   d. Coral Sea
   e. Atlantic Ocean
   f. Mediterranean Sea
   g. Jersey, Guernsey, Alderney, Sark, and several islets lie in the English Channel.
3. a. Asia. At last count, there were fewer than 1,000 wild giant pandas in China, and the number is declining.
   b. Related to the monkey, lemurs are confined to the island of Madagascar.
   c. North America
   d. The giant tortoise (or *galápago* in Spanish) makes its home on the Galápagos Islands.
   e. Asia. The great wildcat is indigenous to India.
4. Peninsula
5. Iraq
6. South America and Africa
7. The Northern Hemisphere
8. You would have to travel northwest, but the cardinal direction is north.
9. France, Italy, Switzerland, Austria, Yugoslavia, or Albania.
10. Siberia is home to more than three-quarters of the world's reindeer, but they can also be found in Canada, Lapland, and Alaska.
11. The Amazon rain forest covers most of Brazil and parts of Colombia, Venezuela, Peru, and Bolivia.
12. Timbuktu, in Mali, remains an important North African trade center.
13. Damascus, Syria
14. The English Channel
15. Austria, Hungary, and Yugoslavia
16. a. The Bering Strait
    b. Canada
    c. It's a Russian version of the Aleutian word *Alakshak,* meaning "great lands."
17. a. Asia
    b. China
18. a. East Germany
    b. East Germany
    c. Federal Republic of Germany
19. The Baltic States
20. Czechoslovakia, Romania, Yugoslavia, and Poland
21. The Sorbonne is in Paris, France, so you'd have to be fluent in French.
22. The Atlantic Ocean
23. Our Arabic numerical system was developed in Arabia, a geographic region that spans modern Syria, Iraq, Saudi Arabia, and other countries in the Middle East.
24. Four: Michigan, Huron, Superior, and Erie

United States

Canada

United States

Pacific Ocean

Atlantic Ocean

Cuba

Mexico

Belize
Honduras
Nicaragua
Guatemala
El Salvador
Costa Rica
Panama

NORTH AMERICA

North America comprises more than 9,360,000 square miles in the Western Hemisphere. The world's third largest continent, it is bounded on the north by the Arctic Ocean, on the east by the Atlantic Ocean, on the west by the Pacific Ocean, and on the southeast by the Gulf of Mexico and the Caribbean Sea. Connected to South America by the land bridge that makes up the countries of Central America, North America's land area is shared primarily by three countries: Mexico, Canada, and the United States. The geography ranges from the frigid Arctic region in Alaska and northern Canada to the warm, breezy shores of the Caribbean Sea and the Gulf of Mexico, from frozen plains and grassy prairies to deserts in the southwestern United States and northern Mexico to rain forests in Central America.

The earliest humans in North America probably migrated from Asia, crossing a land bridge now submerged under the Bering Strait, the 53-mile-wide body of water that connects the Arctic and Atlantic oceans and now separates North America from Asia. Spread throughout North America, their descendants call themselves by such tribal names as Inuit, Apache, Navajo, Lakota, Iroquois, Seminole, and Cherokee.

Scandinavians were the first known Europeans to set foot on the North American continent—and may, in fact, have gone as far inland as Minnesota. Norse explorer Eric "the Red" Thorvaldson colonized the island of Greenland probably in the latter half of the tenth century, but for unknown reasons the settlement disappeared. He was followed by his son, Leif Ericsson, who landed on possibly the continental islands of Labrador and Nova Scotia around the year AD 1000 and named the landmass Vinland.

It was the next wave of Europeans, however, who left the most indelible impression on North America, beginning in 1492 with Christopher Columbus's arrival at San Salvador (Watling Island), one of the islands in the Bahamas, an archipelago of some 700 continental islands southeast of Florida. Columbus went on to

discover islands and establish settlements throughout the Caribbean Sea. And in 1497, Giovanni Caboto (known better as John Cabot), a Venetian explorer sailing under the English flag, discovered the North American coastline.

The Spanish established the first settlements on the North American mainland in 1510 along the Gulf of Darién in present-day Panama. In 1519, Spanish conquistador Hernán Cortés ventured onto the shores of Mexico and subjugated the Aztecs, who, along with Central America's Mayas, had built a great civilization, creating impressive monuments and temples still found throughout Mexico.

The Spanish landed in Florida in 1513, during Ponce de Leon's exploration of the coast in his search for the fabled Fountain of Youth. In 1534 the French established Fort Caroline, the first European settlement in the future United States. By the following year, however, the Spanish destroyed the settlement at the mouth of north-

eastern Florida's St. Johns River and founded St. Augustine, now the oldest permanent European settlement in North America.

Although the English didn't set down roots in Jamestown, Virginia, until 1607, their language and culture ended up dominating two of the continent's three major countries, Canada and the United States. The French legacy remains entrenched in the Canadian province of Quebec. (The French language also remains in Haiti, but it is strongly influenced by the Creole culture.) In Mexico and Central America, as well as the Caribbean nations of Cuba and the Dominican Republic, the Spanish heritage culturally connects the people to South America. (The Spanish heritage continues to characterize the culture of Puerto Rico, although the Caribbean island has been politically tied to the United States since 1898.)

Topography

As the discovery of North America progressed, explorers and adventurers found a land of vastly varied geography, ranging from endless prairies and majestic mountains to frozen plains, from deserts to tropical rain forests.

The eastern North American continent is characterized by the Appalachian Highlands, which extend from Alabama to Quebec, New Brunswick, and Newfoundland. The Appalachians began to take shape 400 million years ago;

glaciers during the Pleistocene era carved the ridges and valleys. Today, along the designated Appalachian National Scenic Trail, from Springer Mountain in northern Georgia to Mt. Katahdin in Maine, you find such mountain ranges as the Great Smokies in North Carolina and Tennessee; the Blue Ridge chain through the southern states to Harpers Ferry, West Virginia; Pennsylvania's Alleghenies; and New York's Catskills. Treasure is buried in these hills and beneath their rivers: abundant coal, oil, and gas.

The Appalachians slope into the old, hard rock of the piedmont and then to the soft, sedimentary rock blanketing the coastal plain. Between these two regions lies the fall line, where rivers rush through waterfalls and rapids on their way to the Atlantic Ocean. The fall line has influenced the growth of numerous eastern cities. In 1790, for example, South Carolina moved its capital from the coastal city of Charleston to the piedmont city of Columbia in an effort to unify the state's "Low Country" plantation gentry and the "Up Country" hill and farm folks. Columbia became a literal source of power by harnessing the rushing waters of the Congaree for grain and textile mills and making the owners wealthy. Later, hydroelectric plants replaced the mills. America's earliest roads—and railroads—followed the path of the fall line, clearly identified today from the air by a string of lights linking such fall-line cities as Trenton, Philadelphia, Baltimore, and Richmond.

While the Appalachians characterize eastern North America, having shaped the early history of Canada and the United States, the Rocky Mountains in the West form the continent's backbone. The mountain range dominates the western landscapes of Canada and the United States, from

The Great Lakes

Superior, Michigan, Huron, Erie, and Ontario form the largest body of fresh water in the world. With their connecting waterways, they are the world's largest inland water transportation system. Traveling through them, ships can reach the Atlantic Ocean via the St. Lawrence River. The Illinois Waterway connects Lake Michigan with the Mississippi and thus the Gulf of Mexico.

The U.S.–Canada boundary passes through four of the five lakes. Lake Michigan is the only Great Lake wholly in the United States.

Lake Superior is the world's largest freshwater lake and ranks as the world's second largest lake after the Caspian Sea.

Geofact

For three days each year, a 1,500-square-mile area of Colorado, including the town of Breckinridge, does not belong to the United States. The area was not included in the Louisiana Purchase, the annexation of Texas in 1845, the Mexican land cession in 1848, or the treaty with the Utes (who apparently never laid claim to it anyway). It became part of the United States in 1936, but with the stipulation that it had the right to be a free and independent kingdom three days each year.

the Bering Strait west of Alaska to northern New Mexico. And they form part of a system of parallel mountain chains called a cordillera, stretching from Alaska through South America and into Antarctica.

A ridge of high ground and peaks that runs through the Rockies in Canada and the United States, through Mexico and Central America, and into South America, the Continental Divide is the watershed of North America. Also known as the Great Divide, it separates westward-flowing rivers (emptying into the Pacific Ocean) and eastward-flowing rivers (those that empty into the Atlantic by way of the Gulf of Mexico). The central point is located in Colorado, where many peaks rise more than 14,000 feet, taller than the Swiss Alps.

The Continental Divide becomes a dry central plateau in Mexico, where it covers more than half of the country. The high plateau is bordered on both the east and the west by the Sierra Madre, two mountain ranges formed by the separation of the Rockies. In Central America, the Great Divide runs along the Pacific coast.

The area west of the Rockies ranges from the lush lava-covered Columbia Plateau in southwestern Canada and the northwestern United States to the waterless Great Basin of Nevada and Utah, with its dried-up prehistoric lakes and the Great Salt Lake. Sparse rainfall gives most of the remote, rocky land in the southwestern United States and northern Mexico a desertlike appearance. In southern California's Death Valley, for example, only $1\frac{1}{2}$ inches of rain may fall each year.

In stark contrast, the landscape dramatically changes as you travel up the west coast. Rainfall increases and temperature decreases to provide perfect conditions for the wine-producing valleys in northern California. Beyond them, the Pacific Northwest becomes damp and cool. Fir and cedar trees dominate the cool, damp western slopes of the Cascade Mountains, while pines, grassy meadows, and sagebrush cover the eastern slopes.

The Colorado Plateau, centered around the Four Corners meeting ground of Arizona, New Mexico, Colorado, and Utah, bears testimony to Earth's geological past. It is a spectacular wonderland of sandstone bridges, arches and spires, buttes and mesas, cliffs, ridges, and canyons. The most dramatic sight, of course, is the Grand Canyon, where the Colorado River has carved through the landscape over many millenia.

Forming the eastern perimeter of the Ring of

Fire, the western coast of North America, from Alaska through Central America, experiences greater geological activity—earthquakes and volcanoes—than the rest of the continent. In recent years volcanoes and earthquakes have been reevaluated in terms of tectonic plates colliding or separating. Since the Pacific plate has been moving against the North American plate for millions of years, it is safe to assume it will continue to do so. Indeed, it may not happen for another few million years, but some scientists predict that the land west of the San Andreas fault, including Mexico's Baja California Peninsula, will split from the continent and drift as far north as Alaska's Aleutian Islands. In the meantime, people living near the fault will continue to accept periodic earthquakes as a fact of life.

Between the Rockies and the Appalachians, the vast plains of Canada and the United States are broken into three subdivisions: the Canadian Shield, the Interior Plains, and the Great Plains. Hudson Bay and its surrounding lowlands lie in the center of the Canadian Shield, or Laurentian Highlands, which spreads across most of Canada, including its most populated regions in the southeast. (See map, page 25.) No navigable rivers cut through this land of lakes and swamps, forests and tundra, where the summers are short and the winters are long and bitter. The rugged, infertile terrain has played the principal role in shaping Canadian history and economics. The rocks, after all, are rich in gold, copper, iron, and other minerals. The surface is rich in commercial forests and beautiful lakes and woods as vacationlands.

The Interior Plains extend from the southwest corner of the Northwest Territories and southeast Yukon Territory and across the United States as far south as Arkansas, Texas, and Oklahoma. This area generally includes the American farm belt.

Slowly rising toward the Rocky Mountains, the Great Plains cover Alberta and southeastern Saskatchewan in Canada, and extend southward through the United States and into Mexico. Its extensive sedimentary surface is the result of receding waters following the melting of glaciers. The geography makes it more suitable for cattle and sheep grazing than farming. Rivers have provided the major source of transportation and commerce for most of the history of the continent. Also used to mark boundaries between states and provinces, these bodies of water also served as the magnets for the continent's major settlements.

CANADA

Canada is bounded by the Arctic Ocean on the north, the Atlantic Ocean on the east, and the Pacific Ocean and the United States (Alaska) on the west. Canada's southern border with United States is nearly 4,000 miles long.

Canada Facts

The world's second largest country in area, Canada consists of nearly four million square miles of land.

• Not counting the Great Lakes, which Canada shares with the United States, Canada has 35 lakes that are over 50 miles long and 9 lakes that are over 100 miles long. The largest include Great Bear (12,100 square miles) and Great Slave (11,100 square miles) in the Northwest Territories; Winnipeg (9,400 square miles) and Winnipegosis (2,000 square miles) in Manitoba; Athabaska (3,100 square miles) in northeastern Alberta and northwestern Saskatchewan; and Reindeer (2,570 square miles) in Saskatchewan.

• About 9,000 feet at its highest point, the United States Range, a part of the Rocky Mountains, is actually located in Canada's Northwest Territories.

• Bounded by Manitoba on the southwest, Ontario on the south, and Quebec on the east, Canada's 480-square-mile Hudson Bay is the world's largest bay.

• Manitoulin Island, on the Canadian side of Lake Huron, at 1,068 square miles, is the world's largest island in a lake. The island encloses more than 100 lakes, of which 41.09-square-mile Manitou Lake is the world's largest lake within a lake.

Northernmost point: Cape Columbia, Ellesmere Island, 83° 07' north latitude.

Highest peak: Mt. Logan (19,850 feet), Yukon Territory (second highest peak in North America).

Although the Norsemen arrived on the continent during the ninth century, and John Cabot, sailing under the English flag, discovered Newfoundland and Nova Scotia in 1497, the French pioneered the settlement of Canada. They called it New France and founded Port Royal (1605), Quebec (1608), and Montreal (1642).

In 1713 England began eroding French influence, gaining control of Newfoundland and Acadia, which are known today as Prince Edward Island, New Brunswick, and Nova Scotia. The Acadians refused to maintain neutrality in the conflicts between England and France, and England expelled them in 1755; the French Canadians dispersed throughout the coastal regions farther south, but their greatest influence was felt in Louisiana, where their descendants are known as Cajuns.

By late 1760 the English had gained complete control of Canada, although the population was almost entirely French. The English presence didn't begin to grow until the American Revolution, when American colonists loyal to England moved to Canada.

In 1867 the Dominion of Canada included Upper Canada (Quebec), Lower Canada (Ontario), Nova Scotia, and New Brunswick. Prince Edward Island joined the federation in 1873 and British Columbia in 1871. In 1869 Canada bought the Hudson Bay Company's landholdings in the vast central plains and eventually carved out the provinces of Manitoba (1870), Alberta (1905), and Saskatchewan (1905). Newfoundland became Canada's tenth province in 1949.

Conflicts between the English and French continue even today, as evidenced in the province of Quebec. Although guaranteed the right to maintain their language, religion, and civil law since 1774, Quebecois have spawned a separatist movement that has troubled the Canadian government in recent years. In 1974 Quebec's provincial government defied Canadian law and made French the province's official language. (English is Canada's official language, required in all government and court proceedings.) In 1990 the Canadian government

North American Volcanoes

Lassen Peak (10,453 feet) in California is one of the two active volcanoes in the contiguous United States that volcanologists observe closely; its last activity occurred between 1914 and 1917.

The other most watched volcano, Mt. St. Helens in southwestern Washington, spewed smoke, ash, and debris in 1980.

Other volcanoes include:
Mt. Hood (Oregon)
Mt. Mazama (Oregon)
Mt. Rainier (Washington)
Mt. Baker (Washington), which has been steaming since 1975 but shows no signs of pending eruption.

granted Quebec the constitutional status of "distinct society," but that has done little to diminish the separatist movement.

The border between Canada and the United States is one of the longest friendly borders in the world. A roadside plaque in Milk River, Alberta, reminds travelers that the area was once part of the United States, having been included in the Louisiana Purchase. In 1818, however, England and the United States agreed to accept the forty-ninth parallel as the northern border of the United States across the Great Plains, which includes the border between the state of Montana and the Canadian province of Alberta. The border was extended along the forty-ninth parallel to the Pacific Ocean in 1846 by the Oregon Treaty.

THE UNITED STATES OF AMERICA

The U.S. didn't always look the same as it does today. At first there were just the original 13 colonies. In the early 1800s, Congress authorized the acquisition of additional territories, a process that continued into the mid-twentieth century with the Northern Marianas and Marshall Islands.

The first of these additions was the Louisiana Purchase in 1803. The U.S. government paid France $15 million for 831,321 square miles of land, extending from the Gulf of Mexico to British America (now known as Canada) and from the Mississippi River to the Rocky Mountains.

Florida was acquired in 1822. The 69,866-square-mile tract was purchased from Spain for $5 million. Then came the following:

Texas, 1845. An area of 384,958 square miles was added when the United States annexed the Republic of Texas on July 5, 1845. The territory became a state the following December.

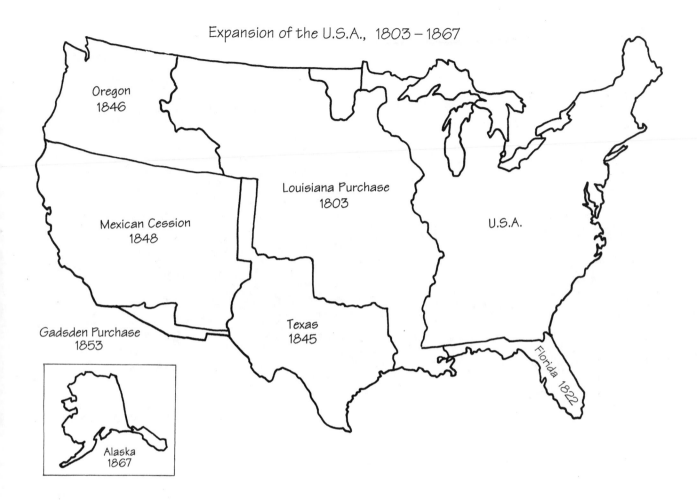

Expansion of the U.S.A., 1803–1867

Oregon
1846

Louisiana Purchase
1803

U.S.A.

Mexican Cession
1848

Gadsden Purchase
1853

Texas
1845

Florida 1822

Alaska
1867

Oregon, 1846. Following the Oregon Treaty, resolving disputes between American settlers and the Hudson Bay Company, England dropped its claim to a 283,439-square-mile area. The treaty extended the border at 49° north latitude to the Pacific Ocean.

Mexican Cession, 1848. When settlers started moving north from Mexico, Texas land came under dispute. President James K. Polk ordered that the land be seized, touching off the Mexican War, which lasted from 1846 to 1848. In February 1848 Mexico signed a treaty agreeing to cede claims to Texas and what are now California, Arizona, New Mexico, Nevada, Utah, and parts of Wyoming and Colorado. The U.S. assumed $3 million in American claims and paid Mexico $15 million.

The U.S.A. during The Civil War

☐ Union States

▨ Confederate States

▨ Slave States That Stayed In The Union

Gadsden Purchase, 1853. Following negotiations by James Gadsden, U.S. minister to Mexico, the United States paid Mexico $10 million for 29,640 acres of land that are now part of New Mexico and Arizona. Texas had claimed sovereignty over this same territory after it won independence from Mexico.

Alaska, 1867. Russia sold 591,004 square miles to the United States for $7.2 million, the result of a treaty negotiated by Secretary of State William Henry Seward. (Although the deal was nicknamed Seward's Folly, gold was discovered almost 30 years later, initiating the famed Klondike Gold Rush.) Alaska eventually gained statehood in January 1959 (and Hawaii would join the Union 7 months later).

U.S.A. Today

In 200 years the borders of states would shift many times. From 1763 to 1767, English astronomers Charles Mason and Jeremiah Dixon surveyed the boundary between Maryland and Pennsylvania to settle the border dispute between the two colonies. The calculations set the border at 39° 43' 11" north latitude. With the 1820 Missouri Compromise, allowing slavery in Missouri but nowhere else west of the Mississippi, the Mason-Dixon Line became known as the dividing line between the free (northern) states and the slave (southern) states.

During the Civil War, the slave states of Delaware, Kentucky, Maryland, and Missouri did not secede from the Union. Western Virginians' resentment toward eastern

Virginians came to a head when Virginia seceded from the Union in 1861; they repudiated Virginia's action and created their own state, originally naming it Kanawha. When it was admitted to the Union in 1863, the state's name was changed to West Virginia.

If Maryland had followed the other slave states and seceded from the Union during the Civil War, the nation's capital would have been surrounded by Confederate states and cut off from the Union states.

The site of the world's first planned capital city, Washington, D.C., was chosen in 1790 and encompassed a 100-square-mile area on the Potomac River. Virginia contributed about 30 percent of the land for the establishment of the capital and Maryland provided 70 percent. Virginia's portion was returned to the state in 1846.

U.S. Highs and Lows

The highest peak (contiguous states): Mt. Whitney (14,494 feet), California. The highest point east of the Mississippi River: Mt. Mitchell (6,684 feet), North Carolina.

Snake River Canyon (Hell's Canyon), on the boundary between Idaho and Oregon, is the world's deepest ravine, at 7,900 feet.

Los Angeles, California, is the world's only major city with a mountain range running through its center.

Washington's Olympia Peninsula is the rainiest place in the contiguous United States, followed closely by southern Louisiana.

The world's shortest river, the D, connects Devil's Lake in Oregon with the Pacific Ocean. At low tide it's only 120 feet long.

Mexico and Central America

By the time of the European explorations in the Western Hemisphere during the fifteenth century, North America's Maya civilization, centered around the Yucatán peninsula in southern Mexico and into Guatemala, had collapsed. Central and southern Mexico was inhabited by the Aztecs, a highly civilized people who founded their capital, Tenochtitlán, in 1325. They probably were descended from an earlier people called the Toltecs.

Christopher Columbus discovered several Caribbean islands, including Cuba and Puerto Rico, from 1493 to 1496. The Spanish founded the first settlement in 1510 along the Gulf of Darién in present-day Panama. Explorers eventually discovered the Yucatán region in 1517 and the coast of Mexico in 1518. Hernán Cortés conquered the Aztecs between 1519 and 1521, establishing Mexico City on the site of the Aztec capital. This became the ruling center for all of New Spain, which eventually extended north, far into the present-day United States (see page 29). The

Spanish subjugated the remaining Mayas in Guatemala and El Salvador. Southern Mexico succumbed to the Spanish invaders in 1524.

In 1821, Spanish Central America (excluding Panama, which joined with Colombia in 1819) was organized under the name Guatemala, but its independence was brief. It joined Mexico in 1822 as the United Provinces of Central America. During 1838 and 1839, the provinces were separated into independent republics that were then briefly reorganized as the Greater Republic of Central America (1895 to 1898).

Although most of its 228,500 square miles of land rise from the Caribbean plate (not from the

North American plate), Central America is geographically considered the southern portion of North America. It extends from the southern border of Mexico to the northern border of Colombia and is bordered by the Pacific Ocean on the southwest and the Caribbean Sea on the northeast. Containing lush tropical rain forests and many active volcanoes, the mountainous region connects the Rocky Mountains and Sierra Madre with South America's Andes.

Folklore has it that Lake Nicaragua contained people-eating sharks, trapped there when the former bay became a freshwater lake.

In 1904 the United States paid the Panamanian government $10 million for perpetual rights to the Panama Canal, with an additional payment of $250,000 per year, a fee steadily increased during subsequent years. In exchange, the United States got the Canal Zone, a 10-mile-wide strip of land across the isthmus of Panama. In the 1970s, the two countries began negotiations for the reversion of the canal to Panama. Two treaties—one

governing the canal's transfer and the other guaranteeing the zone's neutrality after transfer—were ratified by the United States Congress in 1978. They provided for the transfer of the canal on December 31, 1999, when Panama gained full control.

The Río Bravo (called the Rio Grande by its northern neighbors) separates Mexico from the United States. The Sierra Madre, the southern extension of the Rocky Mountains, is divided into two mountain ranges, one on the western (occidental) and the other on the eastern (oriental) side of the Anáhuac Plateau, which covers more than 50 percent of the country and has an average elevation of 6,000 feet.

Mexico's chief rivers are the Balsas, Conchos, Grijalva, Pánuco, Bravo, Santiago, and Usumacinta. Its lakes include Chapala, Cuitzeo, Pátzcuaro, and Texcoco. North America's largest tropical rain forest, Lacandona, once covered 5,000 acres in Chiapas, but more than half of it has been decimated since the early 1970s.

The Continental Islands

Although separated by miles of open sea from what most of us think of as North America, Greenland, Bermuda, the Bahamas, and the islands of the Greater Antilles in the Caribbean Sea are considered part of the continent. These continental islands still reflect the many different cultures of their earliest settlers.

GREENLAND

Less than 17 percent of the 840,000-square-mile Danish island of Greenland is ice-free; the ice sheet covering it reaches a thickness of as much as 11,000 feet. Numerous islands lie around its coastline, which is deeply cut by fjords, narrow inlets of the sea between cliffs and steep slopes. The greater part of the world's second largest island lies within the Arctic Circle.

BERMUDA

Bermuda lies about 640 miles east of North Carolina in the Atlantic Ocean. Also known as the Bermuda Islands or Bermudas (and formerly

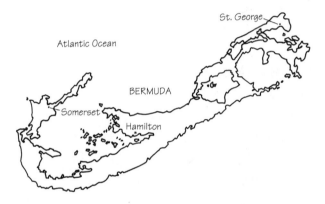

known as the Somers Islands), the British colony comprises some 300 islands, only 20 of them inhabited. The principal island, Bermuda Island, is also known as Great Bermuda and Long Island. The Spanish discovered the islands in 1515 and named them for Juan de Bermudez; the British called them the Somers Islands after Sir George Somers, who landed there on his way to Virginia. In 1815 the capital was moved from St. George's Island to Hamilton, on Bermuda Island. The United States leases military and naval bases on the islands.

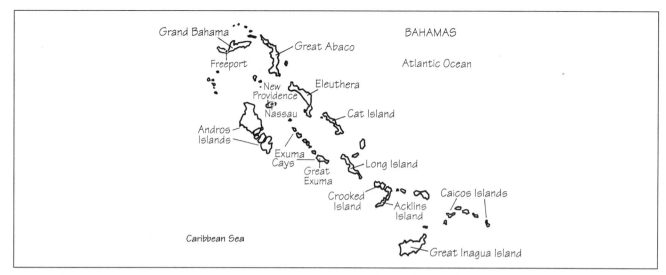

THE BAHAMAS

The name of the Bahamas is from the Spanish *baja mar,* meaning shallow water. An archipelago (or group) of some 700 islands (only 30 of which are inhabited), the Bahamas lie in the Atlantic Ocean, 50 miles southeast of Florida.

Christopher Columbus first set foot on land in the Western Hemisphere at San Salvador (Watling Island) in 1492 and claimed it for Spain. (The actual site of Columbus's landing is still disputed, however.) Only slave raiders and pirates visited the islands until England began settlement in 1647. The British were greatly aided by the influx of British loyalists during the American Revolution. Because of their proximity to the United States, the Bahamas were an important base for Confederate forces attempting to break through the Union's blockade of southern ports during the American Civil War.

The Bahamas became a British colony in 1783 and gained full independence in 1973.

CUBA

The largest and westernmost island in the Caribbean, Cuba lies 90 miles south of Key West, Florida. Low hills and fertile valleys cover more than half the island; its northern coast is steep and rocky, while its southern coast is low and marshy. Sierra Maestra is the highest of three mountain ranges.

Christopher Columbus discovered the island in 1492. At the time, some 50,000 Arawak Indians lived on Cubanacan, the Arawak name for the island, but they quickly died of diseases introduced by the Spanish. By 1511 the Spanish were founding settlements that would become bases for conquistadors to drop off looted treasure, which attracted British and French pirates. A slave-based sugar plantation economy developed during the eighteenth century and enabled Cuba to become a major sugar producer.

Cuba gained its independence from Spain primarily through the intervention of the United States, which went to war with Spain in 1898 following the sinking of the battleship USS *Maine* in Havana harbor. By the terms of the treaty ending the Spanish-American War, Cuba became an independent republic under American protection.

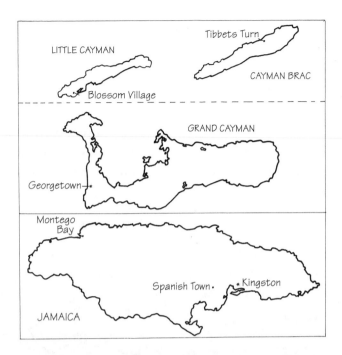

JAMAICA

The 4,400-square-mile island of Jamaica lies 95 miles south of Cuba. It is 145 miles long, with several bays providing good anchorage, including Kingston Harbour and Montego Bay. Volcanic mountains are abundant, with their main ridge running east and west across the island.

Like the other Greater Antilles islands, Jamaica was inhabited by Arawaks when the Spanish began to colonize the island in 1509—and they all quickly died off. It remained under Spanish rule until it became a British possession in 1655. Port Royal was the base of wealthy buccaneers, and the island's capital prospered until it sank into the sea following an earthquake in 1692. Kingston, the new capital founded in 1692, was also severely damaged by an earthquake in 1907.

The island became a British colony in 1866 and gained its independence in 1962.

THE CAYMAN ISLANDS

The Cayman Islands are a cluster of three islands—Grand Cayman (76 square miles), Little Cayman (20 square miles), and Cayman Brac (22 square miles). They lie 200 miles northwest of Jamaica, with Cuba to the north and Mexico to the west. A British dependency, the Cayman Islands became administratively independent of Jamaica in 1962.

HISPANIOLA
Discovered in 1492 by Columbus, who named it La Isla Española, the island of Hispaniola is now shared by two independent nations—Haiti and

the Dominican Republic—whose histories are closely linked, although their people speak different languages.

The western third of Hispaniola became a French possession in 1697. In 1795, Spain ceded the other two-thirds of Hispaniola to France. In 1791 a slave insurrection eventually led to a declaration of independence in 1801. The residents of the eastern two-thirds, still largely Spanish, revolted the following year and captured the city of Santo Domingo, the oldest continuous settle-

ment in the Western Hemisphere and future capital of the Dominican Republic.

On January 1, 1804, the world's first black nation created by former slaves was officially rechristened Haiti, its ancient aboriginal name. Haitian attacks on the rest of the island continued into the second half of the nineteenth century, following the establishment of an independent Dominican Republic in 1844.

PUERTO RICO
Puerto Rico is the smallest of the Greater Antilles islands. The land is mostly mountainous, surrounded by a broken coastal plain. Spain ceded its autonomous colony of Puerto Rico

(along with Cuba and the Pacific islands of the Philippines) to the United States following the Spanish-American War in 1898.

Quiz (answers, p. 39)

1. With New York City as your starting and ending point, you plan to travel to Los Angeles, Dallas, Chicago, New Orleans, Seattle, Reno, Miami, and Baltimore. Plan your itinerary so that you do not pass through any state more than once.

2. In what states would you find the following tourist spots?
   a. Everglades
   b. Yellowstone National Park
   c. Waterton-Glacier International Peace Park
   d. Mt. St. Helens
   e. Mt. McKinley
   f. North American Galápagos
   g. Grand Canyon

3. Where in the United States is it possible to stand in four states at once?

4. What states were carved out of the Louisiana Purchase?

5. What extends more than 600 miles from northwestern California to the Gulf of California?

6. What city lies farther north: Minneapolis or Toronto?

7. To travel from Alaska to the lower 48 contiguous states, you have to pass through a foreign country, Canada. Parts of two other states are similarly isolated from the rest of the country. In what states are these areas located?

8. What states lay claim to part of the Mississippi River?

9. Hudson and Chesapeake are examples of what geographical feature?

10. What country outside the North American continent is closest to the United States?

11. What was the capital of the Aztec empire? By what name is it known today?

12. What major city lies on an island in the St. Lawrence River?

13. On your way from New Mexico to Wyoming, you plan to stop in Denver, Colorado. Routes I-25 and I-70 both cross Colorado. Which one should you travel?

14. Beginning on the west coast, you plan to drive across southern Canada to New Brunswick on the east coast. In what order will you pass through Vancouver, Montreal, Winnipeg, Ottawa, Regina, and Calgary? Through what provinces will you travel?

15. What is the $3/4$-mile-long Goat Island's claim to fame?

16. Which of the United States were originally part of Mexico?

17. What Caribbean city is the oldest permanent European settlement in the Western Hemisphere?

18. Home to 200 Eskimos, 180,000 caribou, and dozens of rare Arctic species, the Arctic

National Wildlife Refuge may contain the biggest undiscovered oil field in the United States. Where is the Arctic Refuge?

19. Where are the following located?
   a. Walla Walla
   b. Wausau
   c. Oshkosh
   d. Moose Jaw
   e. Truth or Consequences
   f. Okefenokee
   g. Tabasco
   h. Tippecanoe

20. Where are the following North American cities located?
   a. Moscow
   b. Rome
   c. Panama City
   d. London
   e. Athens
   f. Paris
   g. Cairo

21. Where is Henry David Thoreau's Walden Pond?

22. Where is the Bay of Pigs?

23. On what Caribbean island does the United States maintain a military base, even though it has no diplomatic relations with the island's government?

24. What is known as the graveyard of the Atlantic?

25. Which of the Great Lakes was known as Gitchigumi to Native Americans?

26. Identify the state names that were derived from the following Native American words:
   a. alibamons
   b. alakshak
   c. illini
   d. mici gama
   e. mici zibi
   f. ouisconsin
   g. tanasi
   h. texias

ANSWERS

1. If you start out by traveling west, your route will be Chicago, Illinois; Seattle, Washington; Reno, Nevada; Los Angeles, California; Dallas, Texas; New Orleans, Louisiana; Miami, Florida; Baltimore, Maryland. If you travel first to Baltimore, the route will be exactly the opposite.

2. a. Florida

b. Depending on where you are in the park, you could be in northwestern Wyoming, southern Montana, or eastern Idaho.

c. Established by Canada and the United States in 1932, the park comprises Glacier National Park in northern Montana and Waterton Lakes National Park, a mountain recreational area in southwestern Alberta.

d. Washington. (The volcano was dormant from 1857 until it erupted in 1980.)

e. The highest mountain in North America is located in Alaska's Denali National Park.

f. Off the coast of British Columbia, south of Alaska, the Queen Charlotte Islands are often called the North American Galápagos because the ecosystem supports a wide variety of species.

g. The gorge in the Colorado River cuts through northwestern Arizona.

3. The "Four Corners" is the meeting point of the Utah, Colorado, New Mexico, and Arizona borders at 37° north latitude, 109° west longitude.

4. Arkansas, Iowa, Missouri, Nebraska, and parts of nine other states — Colorado, Kansas, Louisiana, Minnesota, Montana, Oklahoma, North Dakota, South Dakota, and Wyoming

5. The San Andreas fault

6. Minneapolis

7. The northernmost point of the contiguous United States is a peninsula known as the Northwest Angle or the Red Lake Indian Reservation. Part of Minnesota, it extends from Manitoba, Canada, into Lake of the Woods. To get to the rest of Minnesota, take route 525 to route 308, a distance of 50 miles.

Point Roberts, Washington, is located on a peninsula extending from southwestern British Columbia. To get to the rest of Washington, drive into Canada and take route 5 south.

8. From its headwaters on the Missouri River, in Minnesota, the Mississippi River flows southward to its delta in Louisiana. Along the way, it forms the lower boundary between Minnesota and Wisconsin, the Wisconsin-Iowa and Iowa-Illinois boundaries; the boundaries between Missouri and its border states — Illinois, Kentucky, and Tennessee; the boundary between Arkansas and its neighbors — Tennessee and Mississippi; and the northern section of the Mississippi-Louisiana boundary. Since the river's course is constantly changing, so do state boundaries, with states losing or gaining land.

9. Bay

10. Russia, just across the Bering Strait from Alaska

11. Known today as Mexico City, and the capital of Mexico, Tenochtitlán was founded in 1325.

12. Montreal, in Quebec, Canada

13. East-west highways in the United States are generally even numbered; north-south highways, odd numbered. So you would take I-25 from Wyoming to New Mexico.

14. Vancouver, British Columbia; Calgary, Alberta; Regina, Saskatchewan; Winnipeg, Manitoba; Ottawa, Ontario; and Montreal, Quebec

15. Situated in the Niagara River, just above Niagara Falls, Goat Island divides Niagara Falls into the American Falls and Horseshoe Falls.

16. Texas rebelled in 1835 and proclaimed its independence from Mexico in 1836; the Republic of Texas became the twenty-eighth state in 1845. In 1846 the United States went to war with Mexico, and following its victory in 1848 the government dropped $3 million in claims and paid another $5 million for territory that now makes up California, Utah, Nevada, most of Arizona, and parts of Colorado and Wyoming. In 1853 the United States purchased another tract of land that is now in New Mexico and Arizona.

17. Santo Domingo, the capital of the Dominican Republic, was founded in 1496.

18. The frozen plain is in northeastern Alaska, near the Canadian border, and includes a coastal plain along the Beaufort Sea.

19. a. Washington
   b. Wisconsin
   c. Wisconsin
   d. Saskatchewan
   e. New Mexico. In 1950 the popular radio show offered to hold its tenth-anniversary program in any city or town that would change its name to Truth or Consequences. The people of Hot Springs voted overwhelmingly in favor of the idea and the name change took effect on April Fool's Day, 1950.
   f. The 40-mile-long swamp is located in southeastern Georgia and northeastern Florida.
   g. Southeastern Mexico, one of the country's 31 states
   h. The river in northern Indiana flows into the Wabash River.

20. a. Northwestern Idaho, near the Washington border
   b. There's a Rome, Georgia (55 miles northwest of Atlanta) and a Rome, New York (15 miles northwest of Utica).
   c. Florida, on the Gulf coast
   d. The best-known, and largest, North American city with the name is in Ontario, Canada; there's also a London, Kentucky, and a London, Ohio.
   e. Northeastern Georgia, 60 miles northeast of Atlanta; there's also an Athens, Ohio, and an Athens, Alabama.
   f. There's a city named Paris in Arkansas, Idaho, Illinois, Kentucky, Maine, Missouri, Tennessee, and Texas, as well as Ontario, Canada.
   g. Cairo, Illinois, lies at the confluence of the Ohio and Mississippi rivers; Cairo, Georgia, lies 50 miles south of Albany, in southwestern Georgia.

21. From 1845 to 1847, Thoreau lived on Walden's shores in today's Middlesex County, in northeastern Massachusetts near Concord.

22. Also known as Cochinos Bay, the Bay of Pigs is on the southwest

coast of Las Villas province in west central Cuba. In 1961 it was the landing site of a failed invasion of Cuba by Cuban exiles trained by the United States Central Intelligence Agency (CIA).

23. The United States has continued to lease Guantanamo Bay, on the southeast coast of Cuba, since 1904.

24. The ocean off the long, narrow sandbar known as Cape Hatteras, South Carolina, is treacherous for navigators. Many ships have sunk in the area.

25. Lake Superior

26. a. Alabama, from the name of a tribe in the Creek confederacy

b. Alaska, the Russian version of the Aleutian (or Eskimo) word meaning "peninsula," "great lands," or "land that is not an island"

c. Illinois, from the Algonquin word meaning "warriors" or "men"

d. Michigan, from the Chippewa word meaning "great water," referring to Lake Michigan

e. Mississippi, from the Chippewa word meaning "great river"; the Algonquin word *messipi* also means "great river."

f. Wisconsin, believed to be Chippewa for "grassy place"

g. Tennessee, from the name of Cherokee villages on the Little Tennessee River

h. *Texas* was used by Caddo and other Native American tribes to mean "friend" or "ally." (The Texas state motto, in fact, is "friendship.")

Other state names based on Native American languages include Connecticut, from Mohican and other Algonquin words meaning "long river place"; Dakota (North and South) is a Sioux word meaning "friend" or "ally"; Idaho is a Kiowa Apache term for the Comanche, according to one theory, although it may also be a coined word with an invented meaning ("gem of the mountains"); Kansas, Sioux for "south wind people"; Massachusetts, from the name of a tribe named after "large hill place"; Minnesota, a Sioux word meaning "cloudy water" or "sky-tinted water" for the Minnesota River; Missouri, an Algonquin word meaning "muddy water," referring to the Missouri River; Nebraska, from an Omaha or Otos word meaning "broad water" or "flat river," describing the Platte River; Ohio, an Iroquois word meaning "fine or good river."

Caribbean Sea

Guyana

Suriname

Venezuela

French Guiana

Atlantic Ocean

Colombia

Ecuador

Peru

Brazil

Pacific Ocean

Bolivia

Paraguay

Argentina

Chile

Uruguay

42

SOUTH AMERICA

For thousands of years before the arrival of Europeans in the sixteenth century, native communities had been developing in South America. And despite the eventual destruction of these civilizations (primarily by the Spanish), modern South American culture still reflects its Indian heritage. It is estimated that one-fourth of the continent's total indigenous population was spread around South America and consists of four major groups: those along the Caribbean fringe; primitive cultures in the tropical rain forest of the Amazon and other lowlands; and tribes in the Brazilian Highlands and most of southern South America. The Cuzco Basin of the Peruvian Andes, however, was the center for the greatest and most advanced civilization of South America, the Incas.

About 1,000 years ago, a number of cultures thrived in the valleys and basins in the Andes and along the Pacific coast. Religions stimulated the construction of houses of worship. Sculpture, painting, and other art forms flourished. The llama was domesticated as a beast of burden, a source of food, and a producer of wool.

In the twelfth century, the Incas began extending their authority beyond their cultural center in the Cuzco Basin. Their achievements are remarkable, particularly compared to those of other ancient civilizations. The Egyptians and even the Aztecs, for example, flourished around rivers and waterways that provided avenues for the exchange of goods and ideas. The Incan empire, however, was forged in high, elongated plateaus, or altiplanos, in the Andes. The altiplanos are separated from each other by some of the world's most rugged terrain, where high, often snowcapped mountains alternate with precipitous canyons.

Eventually the Incas moved southward and conquered the people along the coast of what is now Peru as well as other altiplanos. Integrating the conquered tribes into their empire, they marched onward as far as what are now central Chile and northwestern Argentina. By the time the Europeans arrived, they had conquered tribes in what are now Ecuador, Colombia, and Bolivia. The sheer size of the empire placed inordinate pressure on the stable functioning of the state. The Incas chose to divide it, with Cuzco administering the south and Quito overseeing the rebellious north.

At its height, the population of the Incan empire may have numbered about 25 million. Just as the Aztec empire was ripe for internal revolt when the Spaniards arrived in Mexico, the

European entry into western South America was equally timely. But while the Incan empire easily disintegrated in the face of the Spanish invasion, the civilization's social values remained a part of Indian life, creating a division between descendants of the Iberians (Spanish and Portuguese) and Indian populations. Even today the Incas' official language, Quechua, is still spoken by millions of Indians living in Peru, Ecuador, and Bolivia.

The location of native populations had a considerable impact on the thrust of the European invasion. The Incas were rich, with gold and silver, productive farmlands, and a sizable work force. So the Spanish began conquering areas in northwestern South America—now Peru, Ecuador, and Colombia. Before long Lima, founded by Francisco Pizarro in 1535, was one of the richest cities in the world, its wealth accumulated through the destruction of the Incan empire.

Today the population of South America remains concentrated on the periphery of the continent—in large part because of two massive geographical features: the Andes and the tropical rain forest of the Amazon River basin. The Andes

extend for a total of 4,500 miles, from the southern tip of South America all the way up the west coast. (As pointed out previously, the Central American mountains connect the Andes to North America's Rocky Mountains.) The South American mountains are divided into a number of individual ranges. At the border between Chile and Argentina, the Andes reach their greatest heights, about 23,000 feet. At an elevation of about 12,600 feet, La Cumbre, also known as Uspallata, is one of the few passes through this mighty wall.

In Bolivia the mountains cover two-fifths of the country and enclose lakes Poopó and Titicaca. The Andes then turn northwest and extend the length of Peru, in many places covering areas 200 to 300 miles wide and reaching heights of more than 22,000 feet. The system narrows in Ecuador, where the greatest peaks are lower than 20,600 feet. In southern Colombia, it separates into three ranges, with peaks less than 19,000 feet.

South America's Four Regions

South America can be viewed as four regions:
• The Caribbean North—Venezuela and Colombia, along with the Guianas—with its tropical plantations and history of African slavery and indentured servants from India.
• The Andean West—Ecuador, Bolivia, Paraguay, and Peru—dominated by the Andes and populat-

44

ed with descendants of the Indians.

- The South—consisting of Argentina, Chile, and Uruguay.
- Brazil—dominating half the continent, with its tropical rain forest covering more than half of the country.

The Caribbean North

Colombia, Venezuela, and the Guianas are the only countries in South America that are bound by the Caribbean Sea. The tropical plantations and a history of African slavery and indentured servants set the Caribbean North apart from the rest of the continent.

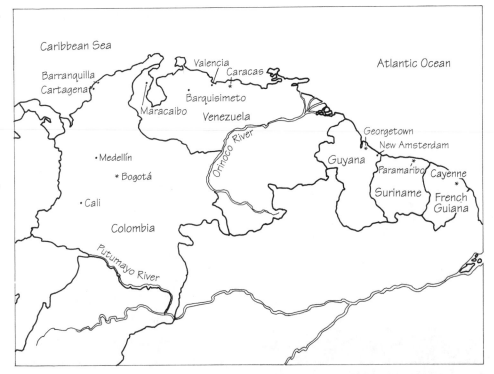

COLOMBIA

At the northwestern corner of South America, Colombia comprises nearly 440,000 square miles. It is bounded on the north by the Caribbean Sea, on the east by Venezuela and Brazil, on the south by Peru and Ecuador, on the west by the Pacific Ocean, and on the northwest by Panama, the border marking the geographical division between the Americas. Colombia is the only South American country with coasts on both the Pacific Ocean and the Caribbean Sea.

Western and central Colombia are covered by the northern end of the Andes, with the highest peak, Cristóbal Colón, rising more than 19,000 feet. The most densely populated region is the fertile plateau and valley between the central and the eastern cordilleras (a system of mountain ranges often having parallel chains). Eastern Colombia is a jungle-covered plain drained by tributaries of the Amazon and Orinoco rivers and inhabited mostly by isolated Indians. Rivers include the Magdalena and the Cauca (a tribu-

tary of the Magdalena), both flowing northward between cordilleras; in the south, the Putumayo forms most of the boundary with Peru.

Colombia mines about half the world's emeralds. Other products include gold, oil, and, of course, coffee.

In the 1530s, following the defeat of the highly civilized native population in the north of South America, Spain established the first settlement, Santa Fé de Bogotá. In 1886 it became the Republic of Colombia, still including Panama, which revolted and finally broke away in 1903.

VENEZUELA

Covering more than 352,000 square miles, Venezuela is bounded on the north by the Caribbean, on the south by Brazil, on the east by Guyana, and on the west by Colombia. The highest mountain range, the Cordillera Mérida, is located in the west and is a northeastern branch of the Andes, with its highest point at Pico Bolívar (16,400 feet). The Cordillera de Venezuela runs along the Caribbean coast in the north, with smaller ranges along the southern border with Brazil. The mountains surround the central grassy plains (*llanos*), irrigated by the chief river, the Orinoco, and its tributaries. The Orinoco system practically covers the entire country and has a thickly wooded delta. Lakes include Lake Maracaibo, notable for its petroleum products, and Lake Valencia. Angel Falls in southeast Venezuela is the world's highest waterfall.

Rich in oil, Venezuela is one of the five founding members of OPEC (Organization of Petroleum Exporting Countries). It also produces diamonds and gold.

Its name meaning "little Venice," Venezuela was one of the first South American colonies to revolt against Spain but

Geofacts

At more than 3,200 feet, Venezuela's Angel Falls is the world's highest waterfall. But the Iguaçú Falls, on the border of Argentina and Brazil, is the mightiest. The falls are divided by forested islands along 2 $\frac{1}{2}$ miles. There are more than 20 cataracts averaging a height of 220 feet.

Although we tend to think of South America as being west of North America, virtually the entire continent lies farther east than Miami, Florida.

didn't win its independence until 1821. It was initially federated as the Republic of Gran Colombia, even though the South American liberator Simón Bolívar was born in Venezuela. Venezuela has long claimed sovereignty over more than half of Guyana's territory in the east.

THE GUIANAS

In "Latin" South America, the Guiana region, with its Dutch, British, and French background, is an anomaly. Covering about 690,000 square miles in northern South America, it falls between the Atlantic Ocean and the Orinoco, Negro, and Amazon rivers, and includes Guyana, Suriname, and French Guiana. (Geographically, it also includes southern and eastern Venezuela and northern Brazil.) The Guianas are major sources of bauxite, used in the production of aluminum.

GUYANA

Occupying 83,000 square miles on South America's northern Atlantic coast, Guyana is bordered on the north by the Atlantic Ocean, on the east and southeast by Suriname, on the south by Brazil, and on the west by Brazil and Venezuela. Guyana's coastal region, where most of the population lives, is predominantly low-lying marsh, providing rich soil for agriculture; inland, grassy savannah slopes up to the mountain ranges of the west (the Pakaraima) and south (the Guiana Highlands). The highest peak is Roraima (9,200 feet) in the Pakaraima Mountains, near the Guyana-Venezuela-Brazil boundary.

Guyana became a Dutch colony in the early sixteenth century. In the late eighteenth century the British captured it and founded Georgetown. During the early part of the nineteenth century, however, the colony went back and forth between the Dutch and the English until the English finally were victorious and in 1831 united Georgetown and the Dutch settlements as the crown colony of British Guiana. The colony gained its independence in 1966 and returned to its traditional name of Guyana.

Guyana's boundaries with its neighbors have long been a source of contention. British claims on the western borders were upheld through arbitration in 1899, although Venezuela continued to claim sovereignty over western Guyana. While the boundary with Brazil was arbitrated in 1904, the border with Suriname remains unresolved.

SURINAME

Suriname (63,250 square miles in northern South America) is bounded on the north by the Atlantic Ocean, on the east by French Guiana, on the south by Brazil, and on the west by Guyana. Southern and central Suriname is a forested plateau and extensive savannah. Mountains in southern and central regions have peaks only as high as 4,200 feet; the Tumuc-Humac Mountains along the Brazilian border are as high as 3,000 feet. The southern region is almost entirely unexplored. The capital is situated on the Suriname River in the north.

The English established the first settlements on the Suriname River in 1651 but transferred the colony to the Netherlands in 1667 in exchange for the Netherlands' New Amsterdam, present-day Manhattan (New York City). Following the abolition of slavery in 1863, laborers were imported from British India and the Dutch East Indies. Having been incorporated into the Kingdom of the Netherlands and achieving autonomous rule in the mid-twentieth century, Suriname gained full independence in 1975. Independence was opposed by the East Indian population, and in the months before independence became official, about 40 percent of the population (mostly East Indians) emigrated to the Netherlands.

FRENCH GUIANA

North and west of Brazil and east of Suriname, French Guiana, comprising more than 35,000 square miles, lies on the northeastern coast of South America. It is divided into two regions—the coastal Cayenne, and the forested hinterlands that cover most of the country. Like Suriname, French Guiana is separated from Brazil by the Tumuc-Humac Mountains in the south. The Maroni River forms the border with Suriname,

and the Oyapock River forms the boundary with Brazil in the east.

The French established the first settlement in the early 1600s but neglected the colony until the eighteenth century. Beginning with their opening in the mid-nineteenth century, Devil's Island and other penal colonies hindered development. French Guiana is the only country on the South American mainland that is not independent.

The Andean West

In the west are South America's only two landlocked countries, Bolivia and Paraguay. The region has a large Indian population—50 percent of the population in Bolivia, Ecuador, and Peru, and about 90 percent in Paraguay. The countries are South America's least urbanized; Lima is the only capital that ranks with capitals in other regions as a major urban center.

BOLIVIA

Covering about 424,000 square miles in the central Andes, Bolivia is surrounded by Brazil on the north and east, Paraguay on the southeast, Argentina on the south, and Chile and Peru on the west. The Amazon and Plata river systems drain the hot, fertile Amazon-Chaco lowlands in eastern Bolivia. The east-central region has semitropical forests. More than two-thirds of the country's population live in the central plateau, which

The Incas conquered the Aymarás, with their high pre-Incan culture, in the fifteenth century. The Spanish, in turn, conquered the Incas and reduced them to slaves in the first half of the sixteenth century. Named after South America's great liberator, Simón Bolívar, Bolivia won its independence from Spain in 1825—and since that time has been marked by internal strife. The country has so far experienced more than 60 revolutions, 70 presidents, and 11 constitutions.

ECUADOR

With an area of nearly 109,500 square miles in northwestern South America, Ecuador is bordered by Colombia on the north, Peru on the east and south, and the Pacific Ocean on the west. Two ranges of the Andes, traversing from north to south, divide the country into three zones: hot, humid lowlands on the coasts; temperate highlands between the two ranges; and rainy, tropical lowlands in the east. The highest peaks—Chimborazo (20,560 feet), Cotopaxi (19,350 feet), and Cayambe (19,000 feet)—are all volcanoes. The country is frequently subject to volcanic activity and earthquakes. Short streams flowing to the Pacific include the Esmeraldas and the Guayas, which flows to the Gulf of Guayaquil in the southwest. Streams on the eastern side of the Andes, tributaries of the Amazon and of its headstream the Marañón, include the Napo and its tributary the Curaray. The Archipiélago Colón

has an average elevation of 12,000 feet and is surrounded by the Andes. The highest peaks are Sajama (21,400 feet), Illimani (21,200 feet), and Sorata (21,000 feet), three of the highest in South America. On the Peruvian border, Lake Titicaca is the world's highest navigable lake and was once at the center of early South American civilizations; islands in the lake hold ruins of the ancient Incas. Bolivia's tin deposits are among the richest in the world.

Geofacts

La Paz is the highest capital city in the world. At an elevation of 12,001 feet (more than 2 $^1/_2$ miles above sea level), the air in the Bolivian capital is 35% thinner than it is at sea level. It is difficult to start a fire because there is so little oxygen. Food cooks slower; cocktails have more kick. And people reared there develop abnormally large lungs. (At more than 9,200 feet above sea level, Quito, Ecuador, is the second highest capital city in the Western Hemisphere.)

Ruins of the Incas' sacred city, Machu Picchu, sit on a mountaintop some 8,000 feet above sea level overlooking Peru's Urubamba River. Stone houses possess trapezoidal doorways, and the precision-crafted buildings display the neat lines, beveled edges, and mortarless seams that characterized the best Incan architecture.

(the Galápagos Islands), 600 miles off the west coast, became the possession of Ecuador in the early nineteenth century.

Along with Venezuela, Ecuador is one of the two South American members of OPEC.

Native Americans in the northern highlands formed the kingdom of Quito around AD 1000. Before the advent of the Spanish, however, it had been absorbed through conquest and marriage into the Incan empire of modern Peru. Since its independence from New Granada, Ecuador's borders have been a major source of friction with its neighbors. Its boundary with Colombia was settled in 1916, but its boundary with Peru—despite three treaties in the latter half of the nineteenth century—was not settled until 1942, when a larger part of the region between the Marañón and Putumayo rivers was assigned to Peru.

PARAGUAY

One of only two countries in South America that are landlocked, Paraguay is bounded on the east by Brazil, on the south and west by Argentina, and on the north by Bolivia. Its nearly 157,000 square miles are sliced by the Paraguay River. The western half of the country, in the fertile Gran Chaco region, is mostly low plains, jungle, and a lot of swampland; the eastern half is mostly densely forested ridges, fertile plains, and grasslands lying between the Paraguay and Paraná rivers. Unlike most of the continent's mountainous neighbors, Paraguay is only 2,200 feet at its highest point. In addition to the Paraguay and Paraná rivers, the Pilcomayo, which is a tributary of the Paraguay, forms part of the boundary with Argentina. The 100-square-mile Lake Ypoá lies in southern Paraguay, near the Paraguay River.

Asunción was established in 1538, with Jesuit missionaries working among the native Guariní Indians until the priests were thrown out by the Spanish. Plagued by dictatorships since becoming an independent republic in 1811, Paraguay entered a disastrous five-year war (1865-1870) with Bolivia, Argentina, and Uruguay. Provoked by a Paraguayan dictator, the war cost the country 55,000 square miles of the fertile Chaco Valley and practically its entire male population.

Oil exploration, backed by United States companies, was fruitless, but the Itaipu Dam, begun in 1978 on the Paraná River, is said to be the largest hydroelectric generator in the world. The joint Paraguayan-Brazilian effort could produce 12.6 megawatts of electricity, more than even the Grand Coulee Dam in the United States.

PERU

Peru's 496,200 square miles on the Pacific coast are bounded by Ecuador on the north, Colombia on the northeast, Brazil on the east, Bolivia on the southeast, Chile on the south, and the Pacific Ocean on the west. Extending for 1,300 miles in western South America, the country is divided by the Andes into three sharply different zones: the coastal region on the Pacific coast, much of it arid and ranging from 40 to 100 miles wide; the mountain zone, with peaks more than 20,000 feet high; and the centrally located plateaus and deep valleys. The highest peak is Huascarán (22,200

feet). The volcanoes El Misti and Yucamami threaten the south. Beyond the mountains in the east lies the densely forested Amazon plain, watered by the Amazon's tributaries. Rivers include the Marañón in northern Peru and the Putumayo, which forms the boundary with Colombia. The principal lake, Titicaca, is shared with Bolivia. The country is rich in silver and gold, with some oil production.

Peru was the seat of the Incan empire, which had its capital at Cuzco and also ruled Quito (Ecuador) and parts of modern Bolivia and Chile. The Spanish arrived in Peru in 1522 and founded Lima in 1535. The viceroyalty of Peru included Panama and all of Spanish South America except Venezuela until the eighteenth century. Peru declared its independence from Spain in 1821 but didn't achieve final freedom until 1824.

The South

The southern reaches of South America feature the grassy, flat country known as the pampas and the mineral-rich Atacama Desert. Uruguay, Chile, and Argentina have stronger European identities than the other South American countries.

ARGENTINA

With more than 1,072,000 square miles and measuring four times the size of Texas, Argentina is the second largest country in South America. It is bordered by Chile on the west, Bolivia and Paraguay on the north, and Brazil, Uruguay, and the Atlantic Ocean on the east. The country's plain rises from the Atlantic to the Andes on the Chilean border. Mt. Aconcagua (22,800 feet) is the highest point in the Western Hemisphere and the highest peak outside Asia. The northern part of the country is characterized by swamp and the heavily wooded area called Gran Chaco.

To the south, fertile pampas are rich for agriculture and grazing; in the far south, most of the land is composed of arid steppes with some fertile areas in Tierra del Fuego and Patagonia. Principal rivers include the Río de la Plata. Many lakes on the slopes of the Andes are resort areas. Argentina is rich in oil and natural gas.

Nomadic Indians were roaming the pampas when Juan Díaz de Solís discovered the Río de la Plata in 1516. By 1580 the Spaniards had settled Buenos Aires, Ascensión, and Santa Fe. The country won its independence in 1816.

CHILE

Chile exemplifies what geographers call an "elongated state" whose shape contributes to political, administrative, and economic problems. Bounded by Peru on the north, Argentina and Brazil on the east, and the Pacific Ocean on the west, Chile consists of a narrow 2,600-mile-long strip averaging 90 miles wide, 220 miles at its broadest. The Andes in the east serve as a buffer zone against encroachment, and the Pacific coast constitutes an avenue of north-south communications.

One-third of the country is covered by the Andes. Cerro, Aguas Calientes, and other peaks in the north are greater than 19,000 feet; others in the east, along the boundaries with Bolivia and Argentina, are even higher. The mineral-rich Atacama Desert lies in the northern plateau, between low coastal mountains and the Andes. In the center is a heavily populated, 700-mile-long agricultural valley between the coastal plain and the Andes. The Andes cover the entire southern part of the country, with its forests and grazing lands.

South of 42° south latitude, the coast is marked with inlets, islands, and archipelagos. Chief among them is Tierra del Fuego, an archipelago Chile shares with Argentina. Chile and

Chile was controlled by the Incas; in the south were the fierce Araucanians. The Spanish founded Santiago in 1541. Although Chile revolted against Spain in 1810, its independence wasn't assured until 1818. War with Peru from 1836 to 1839 expanded Chilean territory; the War of the Pacific added more Peruvian land as well as the nitrate-rich Bolivian coast to Chile's territory.

URUGUAY

Uruguay's more than 68,000 square miles are bounded by Argentina on the west (and on the south, across the Río de la Plata), by Brazil on the north, and on the east by Brazil and the Atlantic Ocean. The country is composed mostly of rolling, grassy flat country known as the pampas. The Uruguay River, flowing into the La Plata, forms the boundary with Argentina. The country is traversed by the Río Negro, flowing from the northwest to the southeast. The capital and principal city, Montevideo, contains 50 percent of the country's population.

With a large non-Spanish ancestry among its population (about a quarter of its population has Italian ancestry), Uruguay is the most European country in South America, lacking the racial minorities that exist even in Argentina and Chile. Spanish explorers discovered the Río de la Plata in 1516, but no one began settling the area until 1680. Then it was the Portuguese who established the first settlement, Colonia (which can

Argentina also share the largest island in the group, named Tierra del Fuego.

Other island possessions include the Juan Fernández Islands in the South Pacific, about 400 miles west of Chile, and Easter Island, about 2,000 miles west. The country has no significant rivers, but there are numerous lakes in the south-central resort area. Chile is rich in natural sodium nitrate, copper, and oil.

At the time of the Spanish arrival, northern

still be found on maps) and began displacing the native Charrúas Indians.

The Spaniards called the area Banda Oriental (from which came today's official name, Republica Oriental del Uruguay). The Brazilian Portuguese and the Spanish disputed the territory for years, until the Spanish won control. Uruguay gained its independence between 1811 and 1814 but was incorporated with Brazil as Cisplatine Province in 1821. In 1825 it revolted against Brazil, which recognized Uruguay as an independent state in 1828.

Brazil

Brazil's more than 3,200,000 square miles cover nearly half of South America, making Brazil the largest country in the Southern Hemisphere (even bigger than Australia) and the fifth largest country in the world. The country is so big it touches every South American country except Chile and Ecuador. The Brazilian Highlands dominate the eastern and southern parts of the country. The highest point is Pico da Bandeira

(9,495 feet) in the east, on the border between the states of Minas Gerais and Espírito Santo.

Comprising the Paraguay, Uruguay, and Paraná rivers, each forming part of Brazil's boundaries, the Plata river system drains southwestern Brazil. The Iguaçú Falls are found in the Iguaçú River, a tributary of the Paraná. In the east, the São Francisco is navigable for 1,000 miles, its route broken by the 275-foot Paulo Afonso Falls near the river's mouth. Most of Brazil's population inhabits the central and southern parts of the country, producing 75 percent of the farm goods and 80 percent of the industrial products.

More than half of Brazil is covered by the tropical rain forest in the north and central regions, comprising the river basin of the Amazon and its more than 200 tributaries. Other than Yanomamo and other Indians scattered throughout the rain forest, the region is largely unpopulated except in a few scattered cities. The government has historically encouraged the settlement of Brazil's interior, especially in the eighteenth century, when it opened the area to people from São Paulo for the extraction of mineral wealth, particularly gold and diamonds, but also for the planting of sugar and coffee.

The new capital, Brasília, was intentionally positioned on the edge of the state of Goiás to encourage the interior's development. Decisions to move a country's capital have been made elsewhere to underscore changes. In Japan, for example, the government moved the capital from Kyoto to Tokyo—creating the new name out of the same letters—when in 1869 the country entered its modernization age.

The Amazon River basin is a source for many products, including Brazil nuts and vegetable oils, as well as diamonds and other minerals. Brazil is also the world's leading coffee grower.

Brazil is the only Latin American nation deriving its language and culture from Portugal.

In 1500 a Spaniard, Vincente Pinzón, reached Brazil. That same year a Portuguese navigator, Pedro Alvares Cabral, came upon it and named it for the *pau-brasil* wood cargo his vessel carried. While the Spanish were infiltrating and exploiting western South America, the Portuguese were exploring the Atlantic coast. In 1501 the Italian Amerigo Vespucci, for whom the continent is named, led a Portuguese expedition to claim the territory. The Portuguese began their first permanent settlement, São Vicente in São Paulo, in 1532.

Ancient geographers measured longitude by the island of Hierro in the Canary Islands, which they thought was the western limit of the world. By the Treaty of Madrid (1750), which shifted the Papal Line even farther westward, Portugal was granted authority to extend Brazil's boundaries. As a result, not only is Brazil almost as large as all the other South American countries combined, it also possesses nearly half the continent's population.

In the 1700s there were far fewer Indians in eastern South America than west of the Andes (perhaps only one million aborigines inhabited all of Brazil), so when the Portuguese

The Amazon River

The longest river in South America, the 4,080-mile-long Amazon is the world's second longest river and the longest navigable one. Oceangoing ships can sail as far as Iquitos, Peru—2,300 miles from the mouth of the Amazon on the Atlantic. Containing one-fifth of all the fresh water that covers the earth, the Amazon's water volume is greater than that of the Nile, Yangtze, and Mississippi rivers combined. The amount of water pouring out of the Amazon is so great that fresh water fills the open sea for more than 200 miles beyond the river's mouth.

The river's basin in northeastern Brazil drains an area roughly three-fourths the size of the contiguous United States. The world's largest tropical rain forest, the Amazon River basin has been exploited for years. Rubber trees produced large profits, and the Brazilian state of Manaus enjoyed a period of wealth. Then the rubber boom ended around 1910, when rubber plantations elsewhere began producing rubber more cheaply.

Today the Amazon rain forest is bursting with development, with hundreds of legal and illegal miners arriving every year. The forest's plants and trees, as well as its Indians, long isolated from the modern world, are being destroyed.

The effects will be devastating for the entire world. Not only do tropical rain forests affect the global climate, but thousands of species of plants and animals, many still unidentified, provide a source of medicine and possible cures for numerous diseases.

turned to sugar cultivation they began importing slaves from Africa to Brazil's northern and northeastern coasts. As a result, Brazil has South America's largest black population, still largely concentrated in Brazil's northeastern states.

Lesser Antilles

The Virgin, Leeward, and Windward islands form an arc of small islands known as the Lesser Antilles, demarcating the Atlantic Ocean and the Caribbean Sea. Although some islands are now independent, they continue to be administered by former colonial powers—the Netherlands, France, and Great Britain. These European countries fought tenaciously over the islands for centuries and, as a result, left a strong cultural influence on them.

THE VIRGIN ISLANDS

The Virgin Islands (see map, p. 36) are divided between Great Britain and the United States. Officially named the Virgin Islands of the United States, the U.S. Virgin Islands comprise 133 square miles of land, which the United States purchased from Denmark in 1917. They include St. Thomas, St. Croix, and St. John, as well as some 50 tiny islands.

A British dependency, the British Virgin Islands include Tortola, Virgin Gorda, Anegada, Jost-van-Dyke, Peter, and Norman, and about 24 islets. The capital is Road Town, Tortola.

Leeward Islands

Constituting most of the Lesser Antilles, the Leeward Islands are so called because their position shelters them from the prevailing northeasterly winds.

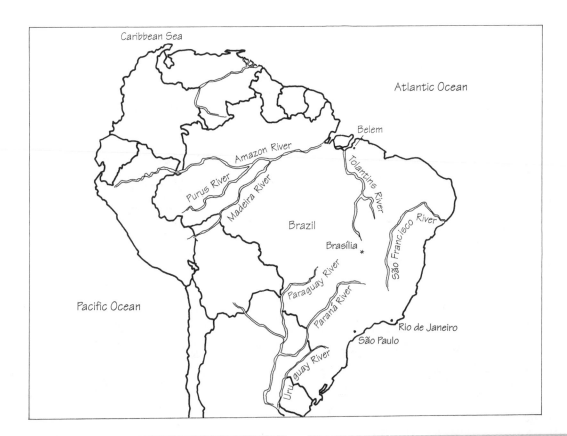

El Dorado

When the Spanish arrived in present-day Colombia, they heard the legend of the Indian nation around Lake Guatavita. Each year their king covered himself in gold dust, took a barge loaded with golden objects out into the middle of the lake, and sacrificed the gold to their god by throwing the objects into the water. The leader then dived in and swam around to wash off the gold dust and sacrifice his "golden skin." He became known as El Dorado, the Golden Man, and his immeasurable golden riches were the Spanish conquistadors' Holy Grail.

Major Volcanoes

There are about 25 active or potentially active volcanoes in Chile and Colombia alone. The world's highest volcano, Guallatiri (19,900 feet), is in Chile. Other major volcanoes include the following:

Chile: Lascar (19,650 feet), world's second highest active volcano
 Tupungatito (18,500 feet)

Colombia: Huila (about 18,900 feet) emits vapors.
 Tolima (about 18,500 feet)
 Puracé (15,600 feet), last erupted in 1949.
 Nevado del Ruiz (17,700 feet), erupted in 1989, causing floods and mud slides that killed more than 22,000 people.

ANGUILLA

Formerly united as part of St. Kitts-Nevis-Anguilla, Anguilla is 35 square miles in area. Its declaration of independence was never recognized by Great Britain, but a truce was signed in 1969 and Anguilla became a British dependency in 1971.

ANTIGUA

Partly volcanic and partly coral formation, with many natural harbors, Antigua covers 171 square miles. Together with Barbuda and Redonda Islands, it constitutes the independent state of Antigua and Barbuda.

GUADELOUPE

Guadeloupe, a department of France, occupies 660 square miles. Its highest point is Soufrière.

MONTSERRAT

Only 40 square miles in area, Montserrat is entirely volcanic. This British dependency has three groups of forested mountains. Plymouth is the capital.

NETHERLANDS ANTILLES

The Netherlands Antilles is an autonomous member of the Netherlands and consists of two island groups—Curaçao and Bonaire near the coast of South America, and St. Maarten, Saba, and St. Eustatius in the Lesser Antilles. Saba (5 square miles) and St. Eustatius (8 square miles), are extinct volcanoes. St. Maarten (13 square

miles) comprises the southern portion of St. Martin, a dependency of Guadeloupe. Curaçao is 83 generally flat square miles. Lying 60 miles north of northwestern Venezuela, it is the largest island in the Netherlands Antilles. The highest point is 1,220 feet. The chief town is Willemstad, the capital of the Netherlands Antilles. The island of Bonaire comprises 111 square miles. It is 30 miles east of Curaçao.

ST. KITTS-NEVIS
St. Kitts (also known as St. Christopher) is 65 square miles in area, and Nevis is 35 square miles. These volcanic islands also include Sombrero. The capital of Basseterre is not the same city as the capital of Guadeloupe.

ST. MARTIN
St. Martin is a 33-square-mile island that has been divided between France and the Netherlands since 1648. St. Martin comprises 20 square miles in the north and is a dependency of Guadeloupe; St. Maarten comprises 13 square miles in the south and is part of the Netherlands Antilles.

Windward Islands
The Windward Islands extend from Dominica in the north to Grenada near the coast of South America.

DOMINICA
The northernmost Windward Island, the republic of Dominica covers 290 square miles. It is volcanic in origin, and its banana plantations were devastated by Hurricane David in 1979.

GRENADA
The southernmost Windward Island, of volcanic origin, Grenada lies about 90 miles north of Venezuela. The 133-square-mile state includes the South Grenadines, tiny islands that include Carriacou and Petit Martinique. In 1983 five members of the Organization of Eastern Caribbean States requested that the United States intervene in the internal affairs of Grenada, following the assassination of Maurice Bishop, Grenada's prime minister and a protégé of Cuba's Fidel Castro, in a military coup. The subsequent invasion included a small military force from Barbados, Dominica, Jamaica, St. Lucia, and St. Vincent.

MARTINIQUE
The 425 square miles of Martinique are dominated by the active volcano Mt. Pelée. The island is the birthplace of France's Empress Josephine, wife of Napoleon Bonaparte.

ST. LUCIA
Occupying 238 square miles, St. Lucia is a volcanic island, similar to its neighbors.

ST. VINCENT

St. Vincent is a string of 600 islets that extend for 60 miles but comprise only 17 square miles. Including the North Grenadines, the area covers 150 square miles. The 133-square-mile St. Vincent is dominated by La Soufrière.

BARBADOS

Although 166-square-mile Barbados is usually considered a Caribbean island, technically it's not; it lies alone in the Atlantic, almost surrounded by coral reefs and with no good harbors. The island is flat, except for hills in the central part. The highest point is Mt. Hillaby (1,104 feet). Barbados is an independent member of the British Commonwealth.

Other Islands

ARUBA

A barren 69 square miles, Aruba lies 18 miles off the coast of Venezuela. The island was separated from the Netherlands Antilles in 1986 and is now an autonomous member of the Netherlands.

TRINIDAD AND TOBAGO

Trinidad is a 1,980-square-mile independent state comprising the island of Trinidad (1,864 square miles) and Tobago (116 square miles). Trinidad is nearly square in shape, with two peninsulas extending from its northwestern and southwestern corners. Spain ceded the island to Great Britain by terms of the Treaty of Amiens in 1802, following the British occupation of the island in 1797. Tobago has changed hands more than any other island in the West Indies, having been held by the Dutch, English, and French, but remained British after 1814. Trinidad and Tobago won independence in 1962 and became a republic in 1976.

Quiz

(answers, p. 63)

1. What group of Pacific islands west of Ecuador inspired Charles Darwin's *The Origin of Species*?

2. Great Britain and Argentina went to war in 1982 over a group of islands in the Atlantic Ocean less than 500 miles northeast of Cape Horn. The Argentines call them the Islas Malvinas. What do the British call them?

3. Match these names with their locales.

a. Indefatigable	Ecuador's name for the Galápagos Islands
b. Pun Run	the world's highest navigable lake, and South America's largest lake
c. Titicaca	mountains in northern South America
d. Archipiélago de Colón	a river in Chile that flows into the Pacific at the city of Concepción
e. Tumuc-Humac	one of the Galápagos
f. Bio-Bio	a lake in central Peru

4. What is a mestizo?

5. What are South American cowboys called?

6. In what former South American capital might you find a "tall and tan and young and lovely girl from Ipanema"?

7. The 1973 film *Papillon* featured Dustin Hoffman and Steve McQueen's escape from the former penal colony on the coast of French Guiana. By what name was the penal colony known?

8. What South American waterfalls were named after Jimmy Angel, who crashed his plane nearby in 1937?

9. Where is the bridge of San Luis Rey, featured in Thornton Wilder's Pulitzer Prize–winning novel?

10. What city does the Trinidad Hilton overlook from a cliff on Belmont Hill?

11. What are the only two landlocked countries in South America?

12. Through which three South American countries does the equator cross?

13. What are the only two South American countries that do not border Brazil?

14. What country took its name from the Incan word for "cold winter"?

15. What country was the only South American oil producer among the five founding members of OPEC (Organization of Petroleum Exporting Countries)?

16. What South American country does Eva "Evita" Perón tell not to cry for her, in the Andrew Lloyd Webber musical?

17. What former colony did the Dutch receive from the English in exchange for New Amsterdam, which became New York?

18. What Caribbean island did United States armed forces invade during the Reagan administration?

19. What country is named for South America's liberator, Simón Bolívar?

20. Juan Valdez, of television commercial fame, is the fictional spokesperson for what country's coffee?

ANSWERS

1. The Galápagos Islands

2. The Falklands. The small population is mainly of British ancestry. England first took control of the islands in 1760 but was ousted by Spain, which later left the islands to Argentina after that country became independent. Argentina claimed the Malvinas and in 1820 revived the settlement of Soledad on East Falkland. British naval forces expelled the Argentines in 1833, but Argentina has never recognized British rule over the Malvinas.

3. Indefatigable is one of fifteen large and many tiny Galápagos islands, together constituting a province of Ecuador called Archipiélago de Colón. Lake Pun Run can be found in central Peru at an elevation of 14,200 feet, but Titicaca, at an elevation of only 12,500 feet, is the world's highest navigable lake and, with an area of 3,200 square miles, South America's largest. The Tumuc-Humac Mountains form the border between Suriname and French Guiana on the north and Brazil on the south. Concepción, Chile, lies at the mouth of the Bio-Bio River, which flows for 238 miles from the Andes to the Pacific Ocean.

4. Mestizos are people of mixed Native American and Spanish or Portuguese blood.

5. Gauchos

6. Rio de Janeiro

7. Devil's Island

8. Angel Falls

9. Peru

10. Port-of-Spain

11. Bolivia and Paraguay

12. Brazil, Colombia, and Ecuador

13. Chile and Ecuador

14. Chile

15. Venezuela

16. Argentina

17. Dutch Guiana, which later became the Republic of Suriname

18. Grenada (pronounced with a long *a*, and not like the "ah" sound of Granada, a city in Spain)

19. Originally known as Upper Peru (despite the fact that it lies southeast of that country), Bolivia was one of Spain's last South American territories to win its independence.

20. Colombia

EUROPE

Prehistoric humans began populating Europe some 12,000 years ago as the glaciers that once covered the entire continent began receding. Ice-covered valleys and frozen plains gradually gave way to green meadows and forests of deciduous trees. And eventually the first great European civilizations began developing in the south, along the Mediterranean shores in and around modern Greece and Italy.

Establishing a civilization that later was known for its art and architecture, politics and government, Greece was influenced, particularly in the areas of astronomy and mathematics, by the Phoenicians and other nomadic Middle Eastern civilizations from Asia's Arabian Peninsula. The Phoenicians, a Semitic people from pre-Israelite Canaan, settled along the coast of modern Syria and developed a maritime culture, which they eventually spread through colonization along the coasts of North Africa and southern Europe. Centuries later, western Europe was also influenced by North Africa's Arab-Berber Moors, who crossed the Mediterranean Sea and invaded Iberia (also known to the Romans as Hispania, today's Spain and Portugal).

The Greeks developed the first great European civilization and contributed to the rise of the Roman Republic in the sixth century BC, followed by the Roman Empire in the first century BC. Whereas the Greeks tended to expand eastward and southward, the Roman Republic, during its height in the second century BC, extended from Europe's British Isles in the northwest to the Persian Gulf along the northeast coast of Saudi Arabia. Today the Roman legacy can still be found in Europe. Many highways, for example, follow the routes laid out by Roman engineers. And Latin, the Romans' language, formed the roots of the Romance languages — modern French, Spanish, Italian, Portuguese, and Romanian.

The Romans' expansion, however, required strong military maneuvering against European tribes to the north, particularly the Germanic

tribes in the frontier beyond the Danube and Rhine rivers. Numerous tribes of Celts, for example, populated much of Europe and were most powerful in the fifth century BC. The Celts, whose culture and religion indicate an Indo-European origin, were first identified in the second millennium BC in the area of modern France and southern Germany known in Roman times as Gaul. They gradually spread to the Iberian Peninsula, the British Isles, and the Balkan Peninsula in southeastern Europe. The Romans (as well as Germanic tribes) finally conquered most of the Celtic territory, but despite the Celts' defeat, their culture survived, most notably in the British Isles.

As the Roman Empire began to decline, hordes of Asian Huns began invading eastern Europe around AD 375, precipitating the migration of Germanic tribes into southern Europe. In the fourth and fifth centuries AD, the Goths and the Vandals from the Baltic Sea region, between the Vistula and Oder river valleys, began invading southern and western Europe. In 429 the Vandals established a kingdom in Africa.

Also in the fifth century, following the fall of the Roman Empire, the Germanic Anglo-Saxons, from Danish shores, and the Vikings, from the Scandinavian Peninsula of modern Norway and Sweden, invaded Celtic Britain. The Vikings also conquered parts of France and Russia, where in 862 their leader, Rurik, became grand prince of Novgorod. In the tenth and eleventh centuries,

the Scandinavian-descended Normans in northwestern France established a kingdom in southern Italy. They also conquered the Anglo-Saxons in Britain, strengthening the still persistent ties between Britain and France.

Meanwhile, the Germanic Allemanni (from the French *allemagne,* which means "Germany") tribe expanded from the lower Rhine valley northward to the Baltic Sea. They were followed by another lower Rhine valley tribe, the Franks, who began spreading throughout central Europe. Early in the ninth century Charlemagne unified the Franks into the Holy Roman Empire, invoking the memory of Rome, and extended Frankish control over much of central Europe and Italy. The Frankish legacy lives on in the names of France and the city of Frankfurt, Germany.

By the end of the twelfth century, with all the maneuvering, Europe had taken on the elements that would shape the modern continent. Geopolitical maps, however, continued to change

even into the nineteenth and twentieth centuries. (See modern geopolitical map, page 64.) In 1867, for example, the Austrian Empire and the Kingdom of Hungary merged into the Austro-Hungarian Empire. This empire, which expanded through annexations, collapsed following its defeat in World War I. This resulted in the creation of new political divisions, including Yugoslavia and Czechoslovakia, uniting various ethnic groups into individual nation-states.

Nation-States

Officially, nation-states first began appearing in the fifteenth century, shortly before Columbus and other explorers began sailing across the Atlantic and finding new lands in the Western Hemisphere. Smaller entities merged into larger units that theoretically included groups of people who spoke the same language, shared similar cultural backgrounds, and generally possessed a common history, as in France, for example. Together they could exercise greater power and authority, as in the unification of Aragon and Castille in what is today part of Spain and France, which led to the expulsion of the Jews as well as the Moors in the south of the Iberian Peninsula and establishment of a strong Catholic state. Nation-states more likely included people who were simply committed to the institutions, law, and politics of the state, as in Switzerland, with its multiple languages, religions, and histories.

The creation of such nation-states was often influenced by Europe's geographical features. (See topographic map, page 75.) The nearly four million square miles of the second smallest continent can generally be divided into 13 regions: the Western Uplands, the British Isles, Iberia, the Lowland Plains, the Central Uplands, the Southern Alpine Region, the Balkans, the Baltic States, the Commonwealth of Independent States, Transcaucasia, the Slavic Nations, Turkestan, and Siberia.

The Western Uplands

SCANDINAVIA

Encompassing the modern countries of Denmark, Norway, and Sweden, Scandinavia was the ancient homeland of the Norsemen. Iceland, first settled by Norwegians, and Finland are sometimes included as Scandinavian countries.

SWEDEN

Sweden and Norway comprise the Scandinavian Peninsula, with the Kingdom of Sweden occupying the eastern and larger section of the peninsula. The southern third of Sweden is lowland with a number of large lakes. The northern part of the boundary between Sweden and Norway is marked by the Kjølen Mountains, which include Sweden's highest peak, Mt. Kebnekaise. The Kjølens are also the source of many rivers that flow southeast to the Gulf of Bothnia.

NORWAY

West of Sweden, the Kingdom of Norway is a mountainous country with extensive plateau regions and many lakes and short streams. Its irregular coastline is characterized by fjords.

DENMARK

Separated from the Scandinavian Peninsula by the Skagerrak and the Kattegat, broad extensions of the North Sea, Denmark is largely a low, flat country with few lakes and no large rivers. The shorelines, however, are indented by many lagoons and fjords, especially in the north and west of the Jutland Peninsula, which includes most of the country's land.

FINLAND

Finland is a land of few hills or mountains. Its highest peak is Haltiatunturi (about 4,300 feet),

on the Norwegian border. Its lakes, however, cover nearly one-tenth of the country's total area of some 130,000 square miles. Swedes conquered the Finns in the twelfth century, and Sweden ceded Finland to Russia in the eighteenth century. It declared its independence in 1917 and after two years of civil war drove out the Russian forces.

The British Isles

The British Isles consist of Great Britain, Ireland, and adjacent islands. Europe's largest island, Great Britain encompasses Scotland, England, and Wales, which along with Northern Ireland in the northeast of Ireland comprise the United Kingdom of Great Britain and Northern Ireland. The Republic of Ireland occupies most of the island of Ireland, known to the Romans as Hibernia.

ENGLAND

Although just 50,000 square miles in area, England has been one of the most important countries in the history of the world. The English people built the first great industrial civilization and founded one of the world's largest empires. In fact, the reach of the Empire (on which, in Victorian times, the British claimed "the sun never set") extended as far as the United States, Canada, Australia, and Hong Kong.

England is the largest of the countries that fall

under the constitutional monarchy of the United Kingdom. The English enjoy the tradition of government and take great pride in the pomp and splendor that are a part of the coronation of their monarch.

England's terrain is mostly rolling land, rising to the uplands of southern Scotland, where the two countries are separated by the granite Cheviot Hills. The landscape is made up of rocky, sand-topped moors in the southwest, downs in the south and southeast, and reclaimed marshes in the eastern central districts. The seas (including the North Atlantic and the English Channel) that wash its shores have protected it from invaders throughout history.

SCOTLAND AND WALES

Comprising nearly 30,000 square miles, Scotland is geographically divided into three regions: the northern highlands, which cover almost two-thirds of the northern part of the country, with Ben Nevis being the highest peak at about 4,400 feet; the central lowlands, which boast the valleys of the Clyde, Tay, and Forth rivers; and the southern uplands, with ranges of hills in which the highest points are as much as 2,700 feet. With numerous islands off the western coast, Scotland's territory also includes three large island groups: Shetland, Orkney, and Hebrides. The coastline is marked by deep inlets known as firths.

An English principality since 1284, Wales occupies a wide peninsula on the southwestern side of Great Britain and covers little more than 8,000 square miles, most of it upland region. At more than 3,500 feet, Snowdon Mountain is the highest point in Wales and England. Wales's irregular coastline is indented by Cardigan Bay in the west and other wide bays.

REPUBLIC OF IRELAND

Geographically, the island of Ireland consists of a central plain with lakes in the northern, central, and western parts, and groups of hills averaging 2,000 to 3,000 feet in the north, south, and west. About 26,600 square miles, the Republic of Ireland occupies the south, central, and northwest of the island and includes the Shannon, the island's chief river and the British Isles' longest, and the highest point, the more than 3,000-foot Carrantuohill in Macgillicuddy Reeks in the southwest.

Having established English rule over a strip of the coast around Dublin in the twelfth century, England subdued all of Ireland in the mid-seventeenth century. In 1922, following civil war,

REGIONS OF EUROPE

Legend:
- Lowland Plains
- Alpine Regions
- Western Uplands
- Central Uplands
- The Balkans
- The Baltics
- Iberia

Atlantic Ocean

Norway
Sweden
Finland
Russia
Estonia
Latvia
Lithuania
Belarus
Ukraine
Moldova

North Sea
Denmark
Netherlands
Great Britain
Ireland
Germany
Belgium
Luxembourg
Poland
Czech Rep.
Slovakia
Hungary
France
Switzerland
Austria
Slovenia
San Marino
Andorra
Monaco
Italy
Croatia
Bosnia & Herzegovina
Serbia
Romania
Bulgaria
Black Sea
Turkey
Macedonia
Albania

Portugal
Spain
Mediterranean Sea
Africa

southern Ireland was named Irish Free State and granted dominion status in the British Commonwealth of Nations. Gradually abandoning its ties with the British crown, it declared itself a sovereign, independent democratic state officially renamed Eire, but remained associated with the Commonwealth of Nations until 1949, when the Republic of Ireland declared itself completely independent.

NORTHERN IRELAND

Formerly six counties in Ireland's Ulster province and consisting of roughly 5,400 square miles, Northern Ireland was separated from the rest of Ireland in 1920, when the region accepted the British offer of Home Rule—to have its own parliament but also to elect members to the House of Commons.

Iberia

Spain and Portugal make up the Iberian Peninsula, or simply Iberia. The greater part of the peninsula is a plateau with an average elevation of 2,000 feet. The Romans called the area Hispania, after the name the Greeks called the people around the Iberus River, now called the Ebro.

PORTUGAL

Portugal's 500-mile coastline affords good harbors only at the mouths of the principal rivers—the Tagus, the Guadiana, and the Douro. Each of these rivers rises in Spain and forms a portion of the Portuguese-Spanish boundary. The country's mountains are also parts of the Iberian Peninsula's east-west ranges. There are no inland lakes, but lagoons can be found at Aveiro and Lisbon at the mouth of the Tagus River.

In ancient times the area now comprising Portugal was inhabited by Lusitanians, who were subjugated by the Romans in the second century BC, conquered by Visigoths from the north in the fifth century, and then by Moors from northern Africa.

SPAIN

Spain occupies the greater part of the Iberian Peninsula, but its coastline affords few good harbors. The Pyrenees separate Spain from France in the north, while the Strait of Gibraltar separates the country from Africa.

The eastern and southern coasts of modern Spain were colonized by Phoenicians and Greeks. The presence of the Pyrenees didn't deter the Vandals from invading the region early in the fifth century. Visigoths ruled the area from 534 to 712, when they were conquered by the Moors, Muslims from North Africa who were finally expelled from the peninsula in 1492.

The Lowland Plains

The Lowland Plains extend from southwestern France northward and eastward across to Poland, along the Gulf of Finland. The plains region also includes a part of southeastern England and the southern tip of Sweden, as well as Luxembourg, Belgium, the Netherlands, and most of Germany.

FRANCE

France is the oldest country in Europe. The city of Paris was a pre-Roman settlement on an island in the Seine River and became the official capital as early as the late ninth century. Comprising nearly 213,000 square miles, France is also the continent's largest country, with coastlines along the Mediterranean Sea, the Atlantic Ocean, and the English Channel, which the French call La Manche. The port of Calais offers an opening on the North Sea, but the major port is Marseilles in the south. Besides the Seine, other chief rivers include the Rhone, Loire, Garonne, and Adour.

Euro Facts

The climate of Europe ranges from the warm shores of the Mediterranean, Aegean, Adriatic, Ligurian, Ionian, and Tyrrhenian seas in the south to the frigid Arctic in the Scandinavian north.

Europe doesn't really constitute a continent at all, but a peninsula of Eurasia, the landmass that includes Europe and Asia. Geographers have set the conventional boundary between Europe and Asia along the Ural Mountains and the Ural River; the eastern banks of the Caspian Sea; the Caucasus Mountains in the south; and the northern coast of the Black Sea, extending along the Sea of Marmara, which slices northwestern Turkey. The continent of Europe is bounded on the west by the Atlantic Ocean, on the north by the Arctic Ocean, and on the south by the Mediterranean Sea.

The cultures reflect the regional variations, as do the languages that can be roughly grouped as Slavic in countries like the Czech Republic, Slovakia, Serbia, and Montenegro; Scandinavian in the northern countries of Sweden, Norway, and Denmark; German in Germany and Austria; the Romance languages of Italy, Spain, France, and Romania; and of course, English.

GERMANY

Germany is generally flat in the north, except for hills that run parallel to the Baltic Sea. It is hilly in the central region and becomes mountainous in the south. The highest peak is Zugspitze, in the Bavarian Alps. Chief rivers include the Rhine, Danube, Elbe, and Oder.

The region east of the Rhine and north of the Danube was inhabited from ancient times by the Teutons, but Germany didn't begin as a political entity until the ninth century. Although the boundaries of modern Germany were never included in the Roman Empire, Germany was the chief component of the Holy Roman Empire, which originated around the year 800, when Charlemagne was crowned Emperor of the West. His reign has sometimes been called the First Reich. An expansionist Germany fought over some regions along its borders for ages. In the northeast, for example, Prussia has gone back and forth between Germany and Poland for centuries.

Following World War II, Germany was partitioned into four occupation zones administered by the United States, Great Britain, France, and the Soviet Union. In 1949 the American, British, and French zones became the Federal Republic of Germany, commonly known as West Germany. The Soviet zone became known as East Germany, or officially the German Democratic Republic (GDR), established as a Communist state in 1949.

Although Berlin was located completely within East Germany, the former capital of Germany was also divided between East and West Germany, with

East Berlin becoming the capital of the GDR. Bonn became the capital of West Germany. In 1990, when Germany was reunited under West German administration, the capital was moved to a reunited Berlin.

BELGIUM

The Kingdom of Belgium is mostly plain, with a wooded hill region known as the Ardennes in the south. Bounded by the North Sea on the northwest, the country has just 42 miles of coastline and comprises only about 11,700 square miles.

In ancient times the Belgae, a people of Celtic descent, inhabited the area that would one day become Belgium. The Belgae were conquered by the Romans and, later, by Germanic peoples. For a period the future Belgium became a part of the territory of the Netherlands and, as a result, eventually came under Spanish control in the sixteenth century. Modern Belgium resulted from the southern Catholic provinces revolting against Spain; the provinces broke away from the northern Protestant provinces in the late sixteenth century. The so-called Spanish Netherlands became Austrian in the eighteenth century. France incorporated the area at the beginning of the nineteenth century. And Holland and the future Belgium were reunited as the independent Kingdom of the Netherlands in 1815. Then in the 1830s Belgium was recognized as an independent kingdom.

THE NETHERLANDS

The Kingdom of the Netherlands comprises more than 14,000 square miles, including about 420 square miles of land reclaimed from the IJsselmeer, a freshwater lake. Nearly a quarter of the country's area is below sea level, with dikes protecting the country along part of the coast. The north-central region was formerly occupied by a large shallow inlet of the North Sea.

The Central Uplands

CZECH REPUBLIC

This mountainous republic became independent in 2003 by cutting its political alliance with the Slovak Socialist Republic, thereby eliminating the nation of Czechoslovakia.

SLOVAKIA

In 2003, the Slovak and Czech regions sundered the bond of nationhood they once shared. Czechoslovakia, the nation they dissolved, was formed from parts of the Austro-Hungarian Empire in 1918.

HUNGARY

Hungary consists mainly of a plain and possesses fertile agricultural land. The country is bisected by the Danube, which flows north to south. Lake Balaton in western Hungary is the largest lake in central Europe.

Europe's Mountains

The Alps dominate southern and central Europe and technically include Italy's Apennines, former Yugoslavia's Dinaric Alps, eastern Europe's Carpathians, and the Pyrenees separating France and Spain, although these are generally considered as independent ranges. The Alpine system extends to the Atlas Mountains in North Africa and eastward into Turkey and beyond.

The Alps' highest point: Mont Blanc (about 15,770 feet) on the border of Switzerland, France, and Italy.

Italy's Apennines extend from the Ligurian Alps in northwestern Italy and run the length of the italian peninsula. They are the source of most of Italy's rivers. Highest peak: Monte Corno (about 9,560 feet).

Eastern Europe's Carpathian Mountains run along the border of the Czech Republic and Poland and extend southward through Ukraine and eastern Romania. Highest peak: Gerlachovka (about 8,700 feet). Romania's 230-mile-long Transylvanian Alps, with their highest peak (Monldoveanul) at more than 8,300 feet, are an extension of the Carpathians.

The Dinaric Alps run parallel to Croatia's Adriatic coast. Highest peak: Voljnac (7,800 feet).

The Balkan Mountains extend across Bulgaria to the Black Sea. Highest peak: Botev (about 7,800 feet).

The Bohemian Forest mountains separate Bavaria (Germany) and Bohemia (the Czech Republic and Slovakia). Highest peak: Grosser Arber (about 4,780 feet) in Bavaria.

Many geographers consider the Caucasus Mountains to lie in both Europe and Asia. Running for about 700 miles between the Black and Caspian seas, the mountains divide the region of Caucasia into Ciscaucasia and Transcaucasia and form part of the boundary between Europe and Asia. There are many peaks greater than 15,000 feet, with Mt. Elbrus (about 18,480 feet) being Europe's highest point.

The Erzgebirge Mountains form the border between eastern Germany and the Czech Republic. Highest peak: Klínovec (more than 4,000 feet).

The Kjølen Mountains form the border between northeastern Norway and northwestern Sweden, in northern Europe's Scandinavian Peninsula. Highest peak: Kebnekaise (about 6,965 feet).

The Rhodope or Rodopi Mountains in southeastern Europe's Balkan Peninsula separate southern Bulgaria and Greek Macedonia. Highest peak: Musala (about 9,600 feet).

The Sudeten Mountains form part of the Polish-Czech border and consist of smaller ranges, including the Riesengebirge and the Eulengebirge. Highest peak: Snézka (about 5,250 feet).

The Ural Mountains run for more than 1,600 miles from the kata Sea to the Kazakhstan, with an average height of 3,000 to 4,000 feet. Highest peak: Mt. Narodnaya (about 6,210 feet). The central Urals, called the Middle Urals, are actually a plateau about 80 miles wide, with an average of 1,000 to 2,000 feet. The Southern Urals consist of three parallel ranges. The range forms part of the boundary between Europe and Asia.

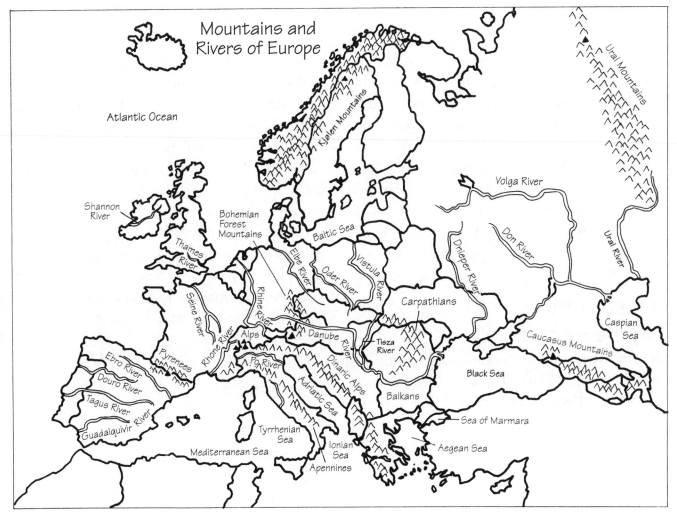

In Hungarian, the country's name (Magyar Nepkoztarsasag or Magyarorszag) is derived from the Magyars, who occupied the area at the end of the ninth century. Following its defeat by the Turks in 1526, most of Hungary was divided between the Turks and Austria. All of Hungary was ceded to the Austrian crown in 1699 and became part of the dual monarchy that made up Austria-Hungary from 1867 to 1918, when it was declared an independent republic following the fall of the Austro-Hungarian Empire.

The capital, Budapest, was created by uniting the cities of Buda, on the right bank of the Danube, and Pest, on the left bank.

POLAND

Poland is essentially a flat country. The area around its southern boundary, however, is mountainous, with the highest peak, Rysy, rising to almost 8,200 feet. Originally a Slavic duchy, Poland emerged between the Oder and Warta rivers in the late tenth century.

Although Poland grew through acquisitions, the country has a history of disastrous wars and political weakness, which has led to the loss of much territory throughout its 1,000 years. Germany didn't finally recognize Poland's western border until 1970.

The Southern Alpine Region

The Southern Alpine Region covers an estimated area of more than 80,000 square miles and extends in a crescent shape for more than 660 miles from the southern coast of France and Italy, northward into Switzerland, eastward into Austria, then southeastward into Serbia and Montenegro.

SWITZERLAND

Switzerland is a mountainous country dominated by the Swiss Alps. Three of its major cities—Bern, the capital; Zurich, the largest city; and Geneva, the site of international conference headquarters—lie in the central plateau. Unlike most other European countries, Switzerland has not one, but four major languages: German in the north-ern—and largest—region; French in the western quarter; Italian in the southeast; and some Romansch in the central southeast. More than half of the population is Protestant, with the remainder predominantly Catholic. While such diversity has been a cause of great turmoil in some countries, Switzerland's stability and unity have inspired international confidence—reflected particularly in the country's banking and insurance industry, which, along with tourism, offsets the necessity to import a substantial part of the country's food and most of the raw materials for its manufacturing industries. No other European country must import virtually all of its raw materials, but in Switzerland the only real natural resource is hydroelectric power.

LIECHTENSTEIN

Formed in 1719 as a principality in the Holy Roman Empire, Liechtenstein comprises 62 square miles along the Rhine, which forms the small nation's boundary with Switzerland. It is also bounded by Austria. The capital is Vaduz.

AUSTRIA

Austria, like Switzerland, is a mountainous country but lies north of the Alps, where it contains many of the Alps' foothills. The Danube enters northern Austria from Bavaria, in Germany, and cuts across to the north to Hungary, with many tributaries sprouting out along the way. Although Neusiedler on the eastern border is the country's largest lake, the western and southern regions boast most of the country's lakes, many of them health resorts.

ITALY

Italy occupies the boot-shaped peninsula that extends southward into the Mediterranean Sea plus the islands of Sicily and Sardinia and a number of small islands. The Apennine Mountains extend the entire length of the peninsula, and the country is bordered in the north, northwest, and northeast by several ranges of the Alps. Many tributaries of the Po, Italy's largest river, form a valley that constitutes the north's great plain.

Poor in mineral resources, particularly coal and iron ore, Italy has substantially decreased its dependence on imported sources of power by developing its hydroelectric power potential. The hydroelectric power market, however, is greater in northern Italy, where, unlike the Mediterranean south, rainfall is less seasonal and the water supply from the Alps is more dependable.

The Balkans

Also called the Balkan states, the Balkans is a collective name for the countries that occupy the Balkan Peninsula—Albania, Bulgaria, Greece, Romania, the former Yugoslavia, and Moldova. Although Turkey lies predominantly in Asia, across the Marmara Sea, its European territory makes up the southeastern part of the peninsula. The Balkan states have often been immersed in warfare among themselves, and as a result, territories have frequently changed hands.

ALBANIA

Albania comprises some 11,000 square miles of very mountainous country, dominated by the North Albanian Alps. The highest peak is more than 9,000 feet.

The Ghegs in the north and the Tosks in the south are ancient Mediterranian peoples who converted from Christianity to Islam. The 1992 elections ended 47 years of Communist rule, but the latter half of the decade saw a quick turnover of presidents and prime ministers. In 1991, however, Albanians began attempting to flee their homeland in the hope of creating better lives. Most of them crossed the Atlantic Sea seeking asylum in Italy.

BULGARIA

The Balkan Mountains cross central Bulgaria, where they are called the Stara Planina, and range from 3,500 to nearly 7,800 feet. However, Mt. Musala, the country's highest point (about 9,600 feet) lies in the Rhodope range in the south. The geography is varied, with plateau, plain, and river valley regions. The Danube River forms most of the country's boundary with Romania.

The ancient Kingdom of Thrace (at one time a region of ancient Greece) corresponded to the region that now comprises central and southern Bulgaria, which derives its modern name from the Bulgars who invaded the region in the sixth century.

GREECE

The culture, art, and traditions of ancient Greece provided the basis for Western civilization. The recorded history dates back to about 3000 BC, when Greece was the center of a powerful and far-flung society. The Greeks were far more advanced than any group that had come before them, bringing to the world the first great dramatists, poets, orators, historians, and philosophers. More than 2,000 years ago, the people of ancient Greece developed a democratic form of government based on the principal that every citizen should take an active part in the state.

Greece is a land of peninsulas. It occupies a main one of irregular shape protruding from the southwestern part of the Balkan Peninsula. Two more large peninsulas extend from this one, and three long spits of land extend into the Ionian Sea. The north-central areas of Greece, Epirus, and western Macedonia are all mountainous.

Mt. Olympus, rising to 9,750 feet in the north near the Aegean Sea, is the highest point in the country. The many island groups include the Ionians off the west coast; the Cyclades on the southeast; and the Dodecanese islands and Euboea, Lésvos, Samos, Khios, and Crete, all in the eastern Aegean Sea.

ROMANIA

Romania was the name taken by the autonomous Danubian Principalities of Moldavia and Walachia when they united in 1861. The country consists largely of fertile plains watered by tributaries of the Danube, with marshlands along the lower Danube and in the river's delta. The southeastern end of the Carpathian Mountains extends into Romania and meets the eastern end of the Transylvanian Alps in central Romania.

YUGOSLAVIA

Croats, Serbs, and Slovenes began pushing southward into the Balkan region during the sixth century and gradually organized into kingdoms that would later make up a large portion of the Socialist Republic of Yugoslavia. The Ottoman Empire conquered all of this territory except Montenegro in the fourteenth and fifteenth centuries. By the beginning of World War I, therefore, Serbia and Montenegro had their independence; Croatia, Dalmatia (later united with Croatia), Bosnia, and Herzegovina belonged to Austria-Hungary; and Macedonia was divided among Serbia, Montenegro, Greece, and Bulgaria.

Following the defeat of the Austro-Hungarian Empire in World War I, the disparate nationalist groups were united as the independent Kingdom of Serbs, Croats, and Slovenes. The name was changed to Yugoslavia during the country's brief period as a monarchy from 1929 to 1932.

Following the establishment of a Communist government after World War II, a new constitution created six constituent republics united under the name Socialist Federal Republic of Yugoslavia. Centuries-old ethnic disputes and territorial conflicts persisted until finally, in the early 1990s, the union broke apart, leaving Yugoslavia to Serbia and Montenegro, now known as the Federal Republic of Yugoslavia.

Serbia and Montenegro

The Federal Republic of Yugoslavia consists of the former Yugoslav republics of Serbia and Montenegro, but Serbian territorial claims extend into the other former Yugoslav republics, including Bosnia-Herzegovina. Conquered by the Ottoman Empire in the fifteenth century, Serbia regained its independence in 1878 and part of Macedonia following the Second Balkan War in 1913.

Having annexed Bosnia and Herzegovina in 1908 over Serbian protests, Austria blamed the hostile anti-Austrian Serbian government for the assassination of Archduke Francis Ferdinand. Austria's subsequent declaration of war on Serbia developed into World War I. The capital is Belgrade.

Bosnia and Herzegovina

Ruled by Croats as early as the tenth century, the Kingdom of Bosnia has been held at various times by Serbians, Hungarians, Turks, and Austrians. Herzegovina had been a principality of the Huns until it was conquered by Bosnia and became an independent duchy in the fourteenth century. Austria reunited Bosnia and Herzegovina. Bosnia and Herzegovina subsequently became part of the Serb-Croat state (later Yugoslavia) in 1918 and then was transformed into the independent Republic of Bosnia and Herzegovina in 1992. The capital is Sarajevo.

Croatia

Although the Kingdom of Croatia was formed in the early tenth century, Croats inhabited the region as early as the seventh century. The boundaries have varied throughout history, at one time extending as far south as the southern borders of Bosnia and Herzegovina.

Croatia's coastal region of Dalmatia features a white limestone plateau known as karst. The Dinaric Alps run along the Adriatic coast, which features good harbors. The capital is Zagreb.

Slovenia

Most of the area now known as Slovenia belonged to Austria until the collapse of the Austro-Hungarian Empire. Settled by the Slovenes in the sixth century and included in the Kingdom of Serbs, Croats, and Slovenes, Slovenia declared its independence from the Republic of Yugoslavia in the early 1990s.

The republic's borders include a tiny stretch of coastal area on the Gulf of Trieste in the northern Adriatic Sea. The Karawanken Alps in the north separate Slovenia from Austria, with the Julian Alps in the northeast bordering Italy.

Moldova

Formerly the Moldavian Soviet Socialist Republic, the Republic of Moldova comprises more than 13,000 square miles, chiefly characterized by the Carpathian Mountains. In 1940 the Soviets formed the Moldavian SSR by merging the 3,200-square-mile Moldavian Autonomous Soviet Socialist Republic and most of Bessarabia, a former Romanian province between the Dniester and Prut rivers.

The Baltic States

The republics of Estonia, Latvia, and Lithuania on the eastern shore of the Baltic Sea were established as independent states in 1917 out of Russia's Baltic provinces, the Polish city of Kovno (Kaunas, Lithuania), and part of Wilno, a department of Poland now known as Vilnius, the capital of Lithuania.

The Commonwealth of Independent States

In 1922 the Soviet republics of Byelorussia (now Belarus), Russia, and Ukraine, and the Transcaucasian Federation (the former union of today's Armenia, Azerbaijan, and Georgia) were organized into the Union of Soviet Socialist Republics under the control of a central Communist government. The Baltic States of Estonia, Latvia, and Lithuania were annexed in 1940.

By 1990, Communism was in its death throes. Eastern European countries were breaking their ties with the Soviet Union, and their Communist governments were ousted from power. One after the other, the 15 Soviet republics declared their independence, despite the vast political and social reforms proposed by then Soviet president Mikhail Gorbachev. Threatened by such reforms, Communist officials dealt the final blow to the Soviet Union with their failed coup in August 1991.

After nearly 70 years of Communism, the Union of Soviet Socialist Republics officially ceased to exist in December 1991, when 11 former Soviet republics constituted themselves as the Commonwealth of Independent States. On the verge of civil war, the Republic of Georgia did not participate in the formation of the Common-

Lithuania
Riga
Tallinn
Vilnius
Estonia
Belarus
Latvia
Minsk
Kiev
Kishinev
Moldova
Ukraine
St. Petersburg
Europe
Asia
Russia
Black Sea
Georgia
Tbilisi
Yerevan
Armenia
Baku
Azerbaijan
Caspian Sea
Aral Sea
Kazakhstan
Turkmenistan
Uzbekistan
Tashkent
Alma Ata
Ashkhabad
Bishkek
Kyrgyzstan
Dushanbe
Tajikistan
Barents Sea
Laptev Sea
Kara Sea

The Commonwealth of Independent States does not unite the republics into a single nation with a central government. Initially formed by the Slavic republics of Belarus (formerly Byelorussia), Russia, and Ukraine, it is a loose association of sovereign nations dedicated, in large part, to reversing the political and economic chaos that developed in recent years. The top governmental body is a council of heads of state and government, assisted by committees of republic ministers in such areas as defense and economics.

The independent states officially recognized as founders of the Commonwealth include the Republic of Armenia, the Azerbaijani Republic, the Republic of Belarus, the Republic of Kazakhstan, the Republic of Kyrgyzstan, the

wealth, although it did send an observer to the negotiations. The Baltic States also chose not to join after they finally gained international recognition of their independence. Estonia, Latvia, and Lithuania did not want any ties with the former Soviet republics.

Republic of Moldova, the Russian Federation, the Republic of Tajikistan, Turkmenistan, the Republic of Uzbekistan, and Ukraine. (Although much of the Commonwealth officially extends into Asia, we are including discussion of all of the republics here.)

Transcaucasia

The countries of Armenia, Azerbaijan, and Georgia make up Transcaucasia, a region dominated by the Caucasus Mountains and bordered by the Caspian and Black seas. Long an area contested by Turks and Russians, Transcaucasia was divided into various Russian provinces and districts at the start of the twentieth century. Following the Russian Revolution in 1917, the region was united in the Transcaucasian Federation. The Federation joined the Union of Soviet Socialist Republics in 1922, the three provinces soon dissolving into three separate Soviet socialist republics.

ARMENIA

Formerly the Armenian Soviet Socialist Republic, the Republic of Armenia occupies 11,500 square miles of southern Transcaucasia, a region dominated by the Caucasus Mountains. Today's Armenia was once part of the ancient country of Armenia lying southeast of the Black Sea and southwest of the Caspian Sea.

AZERBAIJAN

Azerbaijan, or the Azerbaijani Republic, comprises more than 33,400 square miles. The autonomous areas of Nakhichevan and the Nagorno-Karabakh Automous Oblast, which are inhabited by Armenians, are the source of conflicts between Azerbaijan and Armenia.

GEORGIA

The Republic of Georgia occupies mainly the southern slopes of the western and central Caucasus Mountains and the valleys of the Rioni and upper Kura rivers. In 1917 it joined the short-lived Transcaucasian Federation, which was an original member of the Union of Soviet Socialist Republics. Georgia declared its independence from the Soviet Union in 1991 but was not one of the founding members of the Commonwealth of Independent States.

The Slavic Nations

Belarus, Russia, and Ukraine are considered the Commonwealth's three Slavic nations, although other ethnic groups inhabit Russia, which extends eastward through Asia.

BELARUS

The new Republic of Belarus, formerly Byelorussia and sometimes called White Russia, comprises more than 80,000 square miles. For centuries a prize of war, Byelorussia was fought over by Poland and Russia. It was subject to the Poles and the Lithuanians during medieval times and became part of the Soviet Union in 1922. Northern Belarus is crossed by the Western Dvina River; the upper course of the Dnieper flows through the east. Extensive marshes extend along the Pripyat River in the south, and the Bug River forms part of the boundary with Poland. In western Belarus, the upper course of the Niemen and its tributaries flow through land that belonged to Poland until 1921.

RUSSIA

The largest and most powerful of the Commonwealth republics, Russia was the main part of the Russian empire and the first to come under the control of the Soviets in the Bolshevik Revolution of 1917. Formerly the Russian Soviet Federated Socialist Republic, the Russian Federation incorporates nearly 6.6 million square miles of territory that primarily occupies the continent of Asia but also extends into Europe. Its mountains include the entire Ural range in the west and various ranges in eastern Siberia. The great plain of the Volga and Northern Dvina dominates European Russia. Asian Russia possesses the valleys of the Ob', Yenisey, and Amur. Its lakes include Baikal in southern Siberia. The tundra belt stretches across northern Russia and lies on both continents. The northern coastline extends for about 3,000 miles along the Arctic Ocean, the eastern coastline for more than 1,000 miles along the Pacific Ocean.

UKRAINE

The more than 230,000 square miles of Ukraine are chiefly steppe land covered with fertile black soil. The southern border, however, is less fertile clay soil and marshland along the Black Sea. The country is traversed by three great rivers—the Dnieper, Bug, and Donets—and bordered on the southwest by a fourth, the Dniester. Settled by Ukrainians and Ruthenians during the sixth and seventh centuries, its chief town (and now the capital), Kiev, was a leading principality in the Russian empire until its conquest by eastern Asia's Tatars in the thirteenth century. Russia regained the area by bits and pieces beginning in the seventeenth century and eighteenth centuries. As the former Ukraine People's Republic, established in 1917, the country declared its independence from the Soviet Union in 1918, and part of it was conquered by Poland between 1919 and 1938. The Soviet Union reconquered the rest and established the Ukraine Soviet Socialist Republic in 1923.

Turkestan

Kazakhstan, Kyrgyzstan, Tajikistan, Turkmenistan, and Uzbekistan make up the western part of Turkestan, a central Asian region that extends into China and Afghanistan. Between 1920 and 1925, the Soviet Union divided Russian Turkestan into the Kazakh, Kirghiz, Tadzhik, Turkmen, and Uzbek Soviet socialist republics.

KAZAKHSTAN

The more than one million square miles of the Republic of Kazakhstan are characterized by the Kirghiz Steppe in central Kazakhstan, with desert regions in the south. The chief rivers are the Syr Darya in the south, the upper Irtysh in the northeast, and the lower Ural in the west. In eastern Kazakhstan, plateau lands north of Lake Balkhash rise to the Tien Shan and western Altai Mountains on the Chinese border.

KYRGYZSTAN

With the Tien Shan mountain range along the boundary with China and the Altai Mountains in the southwest, the mountainous Republic of Kyrgyzstan comprises more than 76,600 square miles. The mountain peaks range from 16,000 feet to Krygyzstan's highest, Khan-Tengri (23,620 feet), with many glaciers and lakes, including Issyk-Kul, at high elevations. The chief river, the Naryn, is a tributary of the Syr Darya that runs through a high western valley.

TAJIKISTAN

Of all the Commonwealth states, the Republic of Tajikistan boasts the highest peak—the 24,590-feet Communism Peak. Its more than 55,000 square miles are very mountainous, with the Pamirs and the Trans Altai mountain system in eastern Tajikistan. Western Tajikistan is also mountainous, marked by valleys of northern tributaries of the Amu Darya. Russia acquired the area as part of Russian Turkestan in 1895. It became part of the Russian Soviet Federated Socialist Republic in 1924 and the Soviets made it a Soviet socialist republic in 1929.

TURKMENISTAN

Turkmenistan covers more than 188,000 square miles. The western and central regions are level and desert. The east is characterized by plateau. The Amu Darya River forms part of the boundary with Uzbekistan. The Murgab River flows through the southeast.

Turki tribes have inhabited Turkmenistan since the tenth century. Defeated by the Russians, they became part of Russian Turkestan in 1881. The Soviets organized Turkmenistan as a Soviet republic in 1924 and admitted it to the Soviet Union in 1925.

UZBEKISTAN

Consisting mostly of desert and plains regions, Uzbekistan includes the southern half of the Aral Sea and the lower course and delta of the Amu

Darya. The region was originally settled by the Uzbeks, a Muslim people of Turkish origin.

Siberia

Although Siberia lies in Asia, its more than 3.3 million square miles lie mostly in Russia, except for a small part in Kazakhstan. The northern belt along the Arctic Ocean consists of frozen tundra, rich in fur-bearing animals. Low plains, some with extensive marshlands, characterize the west; mountain ranges in the east and southeast; plateaus in southern and central Siberia. Its major rivers, the Lena, Ob', and Yenisey, flow northward to the Arctic. Lake Baikal lies in southern Siberia, north of Mongolia. Following World War II, the Soviet Union implemented large-scale colonization of the region and exploitation of Siberia's natural resources.

Quiz
(answers, p. 88)

1. What body of water laps at the white cliffs of Dover?

2. Where would you find the following landmarks?
    a. the Rock of Gibraltar
    b. Mt. Vesuvius
    c. the Blarney Stone
    d. Trondheim Fjord
    e. the Black Forest
    f. dikes

3. In what country would you be if you were strolling through the famed Tivoli Gardens?

4. What is the world's smallest republic?

5. If you were standing in front of Leonardo da Vinci's painting of *The Last Supper*, in what city would you be?

6. What countries make up the largest island in the British Isles?

7. What is the longest river in the British Isles and the longest in Ireland?

8. In what country would you find Transylvania?

9. In what country were the Allied troops when they landed in Normandy during World War II?

10. Where would you find true Bohemians?

11. Whom would you be visiting if you were lunching at 10 Downing Street in London?

12. If you were at the gambling casino in the commune of Monte Carlo, in what principality would you be?

13. If you were skiing in the Dolomites, in what country would you be?

14. If you took a cooking class at Cordon Bleu in the morning, ate lunch on the Rive Gauche, spent the afternoon at the Louvre, and joined friends for dinner at Maxim's, in what city would you be spending a lot of francs?

15. What French town is famous for its brandy?

16. What do the English call La Manche?

17. Where are the Channel Islands located?

18. Where is the International Court of Justice?

19. What two bodies of water does the Strait of Gibraltar connect?

20. Italy's geographical shape resembles a boot. What island does the boot's foot kick?

21. In what country would you be if you were visiting the Holocaust memorial at Auschwitz?

22. What Swiss city was the birthplace of the International Red Cross (1864) and home of the League of Nations from 1920 to 1946?

23. In February 1945, toward the

end of World War II, U.S. president Franklin Roosevelt, British prime minister Winston Churchill, and Joseph Stalin, the Soviet Union's general secretary of the Communist party, met in Yalta to discuss the disarmament and partition of Germany. Where is Yalta?

24. Following his defeat at Waterloo (in French pronounced vat-er-loo) in 1815, French emperor Napoleon Bonaparte abdicated (for the second time) and spent his last days in exile on the island of St. Helena. Where is Waterloo? Where is St. Helena?

25. His legendary life depicted in the Shakespeare play, Macbeth was the king of what people?

26. Determine the longitude and latitude of the following cities:
   a. Paris
   b. London
   c. Rome
   d. Berlin
   e. Moscow

ANSWERS

1. The Strait of Dover, which connects the English Channel and the North Sea, is generally considered part of the English Channel. The French call the strait Pas de Calais (Calais Passage) after the French city of Calais, which lies directly across from Dover.

2. a. Located in Gibraltar, on the Iberian Peninsula and overlooking the Strait of Gibraltar, the limestone massif (mountain mass) is connected to Spain by a sandy plain.

b. The volcano, which destroyed the ancient city of Pompeii and last erupted in 1944, lies south of Naples, Italy.

c. According to legend, visitors to Blarney Castle, in the Republic of Ireland, are supposed to gain the power of eloquent speech as they hang upside down to kiss the Blarney Stone.

d. Trondheim is one of the many fjords for which Norway is famous.

e. The source of the Danube River, the Black Forest is actually a heavily forested mountain range in western Germany.

f. Much of the Netherlands' coastal area lies below sea level; without man-made dikes, as well as natural coastal dunes, two-fifths of the country would be submerged.

3. Denmark

4. San Marino is a landlocked republic lying within central Italy.

5. Milan, Italy

6. Scotland, Wales, and England

7. The Shannon

8. The region of Transylvania is located in northwestern and central Romania.

9. France

10. Bohemians are the ethnic group that makes up the province of Bohemia in the Czech Republic.

11. The British prime minister

12. Monaco, which is divided into three communes, or districts.

13. One of the Alpine mountain ranges, the Dolomites are located in northeastern Italy.

14. Paris

15. Cognac

16. The English Channel

17. The islands of Jersey, Guernsey, Alderney, and Sark are the largest of the Channel Islands, in the English Channel.

18. International Court of Justice, or World Court, sits in The Hague, Netherlands.

19. The Atlantic Ocean and the Mediterranean Sea

20. The "toe" of the bootlike peninsula that forms Italy faces east, barely touching the island of Sicily.

21. Auschwitz, called Oświęcim in Polish, lies to the west of Krakow, Poland.

22. Geneva

23. The Crimean Peninsula, on the Black Sea

24. Waterloo lies in central Belgium, 12 miles south of Brussels. St. Helena lies in the Atlantic Ocean, about 1,200 miles west of southern Africa.

25. The provincial ruler of Moray, Macbeth became king of the Scots after he killed Duncan I.

26. a. Paris, France

Lat. 48° 48' N, Long. 2° 20' E

b. London, England

Lat. 51° 32' N, Long. 0° 5' W

c. Rome, Italy

Lat. 41° 54' N, Long. 12° 27' E

d. Berlin, Germany

Lat. 52° 30' N, Long. 13° 25' E

e. Moscow, Russia

Lat. 55° 45' N, Long. 37° 36' E

Tunisia

Mediterranean Sea

Canary Islands

Morocco

Algeria

Libya

Egypt

Red Sea

Western Sahara

Mauritania

Mali

Niger

Chad

Sudan

Eritrea

Djibouti

Senegal

Gambia

Burkina Faso

Guinea

Nigeria

Central African Republic

Somalia

Ethiopia

Guinea-Bissau

Côte d'Ivoire

Ghana

Cameroon

Sierra Leone

Liberia

Togo

Benin

Equatorial Guinea

São Tomé & Principe

Gabon

Congo

Zaire

Rwanda

Uganda

Kenya

Burundi

Tanzania

Atlantic Ocean

Angola

Malawi

Zambia

Mozambique

Madagascar

Namibia

Zimbabwe

Botswana

South Africa

Lesotho

Swaziland

Indian Ocean

90

AFRICA

rior to the fifteenth century, when European trade began, Africa was a continent in transition. In central and southern Africa, people were migrating and sometimes struggling for territorial supremacy. Cities were developing in western Africa. A populous kingdom had thrived in Ethiopia, known to have existed in the third century BC. Not only were Africans trading among themselves, but large-scale trading with China, India, Indonesia, and the Arab world had brought foreign crops, customs, and merchandise to East Africans.

Ancient African states included Ghana, which lay northwest of the coastal country that now takes the name. For at least a thousand years before Europeans began exploring Africa, Ghana had a large capital, with markets, religious shrines, and a fortified royal retreat. Taxes were collected and tolls were levied on imports.

Dating to at least the third century BC, Ethiopia is the oldest independent nation in Africa and one of the oldest in the world. The continent's first Christian country, the country once known as Abyssinia, was largely Judaic until much of the population was converted to Christianity during the fourth century. Today in Ethiopia Islam claims slightly more adherents than the Ethiopian Orthodox Church.

Following Ghana's decline, in part a result of the invasion by Muslims from North Africa, cultural, political, and economical power transferred to Mali, centered around Tombouctou (Timbuktu), a thriving center of commerce and trade and one of the world's leading urban centers. Later the state of Songhai arose, centered around Gao, a city that still exists today on the Niger River.

In the south, too, a large state known as Kongo existed for centuries near the mouth of the Zaire River. And in what is today southwestern Nigeria, a number of urban farming communities arose, their walled fortifications affording farmers protection. Surrounding the walled farming communities, the cultivated land could sustain the populations, each numbering in the thousands.

Although the European colonization of Africa didn't actually begin until the second half of the nineteenth century, the Portuguese had

91

Major Islands

Although all of the countries of Africa are now independent, the islands off the continent's coast are still largely owned by non-African nations. Many of these islands also provided haven to the pirates of the Barbary Coast and waypoints for the slave and shipping trades in the seventeenth through the nineteenth centuries.

Off the Atlantic Coast the islands include the following:
- Bioko, owned by Equatorial Guinea
- The Canary Islands (including Tenerife, Fuerteventura, Grand Canary, Arrecife, Hierro, Gomera, and La Palma), owned by Spain
- The Madeira Islands (including Madeira, Porto Santo, and two groups of uninhabited islands), owned by Portugal
- St. Helena, owned by the United Kingdom

Off the Indian Ocean coast the islands include the following:
- Mayotte (a former island in the Comoros Island group), owned by France
- Réunion, owned by France
- Zanzibar, owned by Tanzania
- Madagascar, with nearly 226,700 square miles of area, the world's fifth largest island

begun trading with Africans in the late fifteenth century and were soon followed by other European nations sending vessels to African waters and establishing a string of coastal stations and forts. Even the Dutch settlement on the southern tip of Africa (where Cape Town, South Africa, now lies) was intended only as a resupply station on the months-long voyage to and from the Dutch colonies in Southeast Asia.

Colonialism, however, was encouraged by attempts to abolish the centuries-old slave trade, as well as missionary endeavors and the potential for greater African riches. The partition of Africa that took place during the last three decades of the nineteenth century was conducted by Great Britain, Spain, France, Germany, Portugal, Italy, and Belgium, as well as the Boers, the descendants of Dutch settlers in South Africa. By World War I, only Egypt, Liberia, and Ethiopia remained independent.

The boundaries of most of these countries that achieved their independence during and after the 1960s actually resulted from the partition of Africa in the nineteenth century. Most of Africa's 52 independent nations gained their independence from European colonial powers between 1957 and 1968. Of these countries, 14—Botswana, Burundi, Central African Republic, Chad, Lesotho, Malawi, Mali, Niger, Rwanda, Swaziland, Uganda, former Zaire, Zambia, and Zimbabwe—are landlocked.

The world's second largest continent, Africa encompasses more than 11 million square miles. The coastline, however, is short in relation to the landmass and there are few natural harbors.

Africa is bounded on the north by the Mediterranean Sea and is separated from Europe by the Strait of Gibraltar, between Morocco (Africa) and Spain (Europe). It is bounded on the east by the Red Sea and the Indian Ocean (chiefly the Mozambique Channel and the Gulf of Aden). It is joined on the northeast to Asia at the Sinai Peninsula.

The Role of the Slave Trade

The African slave trade dates back to ancient times, when slaves were sent across the Sahara Desert and were traded throughout the Mediterranean by the Phoenicians. Greek and Roman trade profited from East African slaves in Egypt and the Middle East. Long before the arrival of the Europeans, African middlemen on the east coast raided the interior for slaves and sold them to the Arabs. The slaves were packed by the hundreds into slave ships and carried off to Arabia, Persia, and India.

The large-scale slave trade didn't occur in West Africa until the arrival of the Portuguese and the development of labor-intensive plantations in the Americas. From their coastal stations, Europeans traded with African middlemen, who exchanged slaves for gold, ivory, and spices. Soon the activity was occurring in the foreigners' stations on the coast; coastal African states rose to power by capturing slaves in the interior and selling them to the white men on the coast. The states of Dahomey (now called Benin) and Benin (now part of Nigeria) were built on slave trade. Between 1650 and 1850, some 12.5 million Africans were traded into slavery and many more died in slave-trade wars. The slave trade destroyed families as well as entire villages and cultures.

The continent can roughly be divided into five regions: North Africa, West Africa, East Africa, Equatorial Africa, and Southern Africa.

North Africa

By 30 BC, when the Roman Empire's colonies extended across the northern coast of Africa, Romans were already referring to the area as North Africa. Today Morocco, Algeria, Tunisia, and Libya are still collectively known by that name. Egypt can also be included among these countries, but with a foothold in southwestern Asia, it is more often identified with the countries of the Middle East. Indeed, all of Arab-Muslim North Africa strongly identifies in culture and religion with the Middle East. For this reason Africa itself is often said to begin south of these countries.

The African coastal region, from Egypt to the Atlantic Ocean, was once known as the Barbary Coast, a name derived from its oldest inhabitants, the Berbers. After conquering the region in the seventh century, the Muslims broke it up into the Muslim states of Morocco, Algeria, Tunis (now northern Tunisia), and Tripoli (now northwestern Libya). The Barbary states later

Mountain Ranges

• **Atlas:** located in northwestern Africa; runs from the Atlantic Ocean to Tunisia's Mediterranean coast. Highest peak: Mt. Toubkal (13,670 feet) in Morocco

• **Ethiopian Highlands:** located in eastern Africa. Highest peaks:
Ras Dashan (15,157 feet)
Mt. Batu (14,130 feet)
Mt. Gughe (13,780 feet)

• **Ruwenzori:** located in central Africa between Lake Albert and Lake Edward; the boundary between Uganda and Zaire. The central peak, Mt. Stanley, known in Zaire as Mt. Ngaliema, has two summits: Mt. Margherita (16,760 feet) and Mt. Alexandra (16,750 feet)

• **Ahaggar:** located in southern Algeria. Highest peak: Tahat (9,850 feet)

• **Tibesti:** located in northwestern Chad. Highest peak: Emi Koussi (11,200 feet)

• **Mt. Kilimanjaro (19,340 feet)** in Tanzania is the world's highest mountain that is not part of a mountain range.

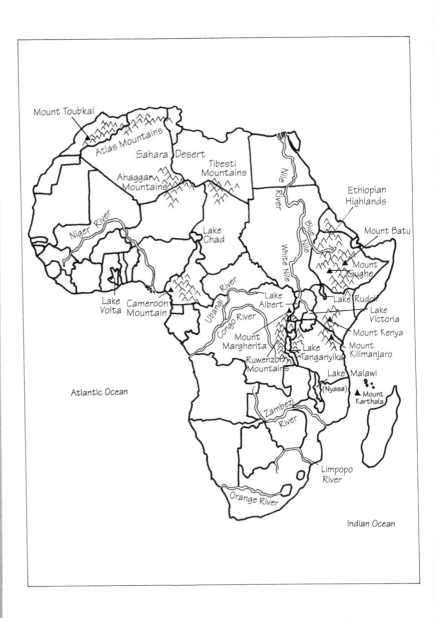

engaged in piracy, continuing to raid on European and American commercial ships until the early nineteenth century.

The Maghreb is yet another name often applied to the region. The term comes from an Arab word meaning "western isle," suggesting the image of the Atlas Mountains as a vast island rising from the Mediterranean Sea and the Sahara Desert.

ALGERIA

The Atlas Mountains and the Sahara Desert traverse the northern region of the Democratic and Popular Republic of Algeria. The Sahara dominates central and southern Algeria, where the Ahaggar (or Hoggar) Mountains form a high plateau. Romans knew the region as Numidia, an ancient country that flourished until its invasion by Europe's Vandals in the fifth century.

EGYPT

The Arab Republic of Egypt is a land dominated by barren, harsh desert. The Eastern Arabian Desert lies between the Nile River and the Red Sea; the greater part of the Libyan Desert (or Eastern Sahara) lies west of the Nile. Since less than 14,000 square miles of the country are actual-ly habitable, nearly all Egyptians live near the shores of the Nile or its tributaries.

Begun at the end of the nineteenth century, the Aswan Dam controls the Nile's seasonal flood waters, enabling Egyptians to irrigate farmlands all year long and grow more than one crop per year. Dedicated in 1971, it created Lake Nasser, one of the world's largest artificial lakes.

Egypt's boundaries incorporate the Sinai Peninsula, giving the country a foothold on the continent of Asia. Recently Israel has contested, militarily and politically, Egypt's Asian territory. In 1958, Egypt briefly joined with Syria to form the United Arab Republic, which was dissolved in 1961.

LIBYA

Libya, or the Socialist People's Libyan Arab Jamahiriya, takes its name from the ancient Greek word for "North Africa outside Egypt."

Largely desert, the country basically occupies the northeastern section of the Sahara, called Fezzan in southwestern Libya, and the Libyan Desert in eastern Libya. The Hamada el Homra desert plateau lies in the northwest. Despite the predominance of desert, agricultural areas can be found along the coast, in the country's northeastern Barqa Plateau, and in the oases.

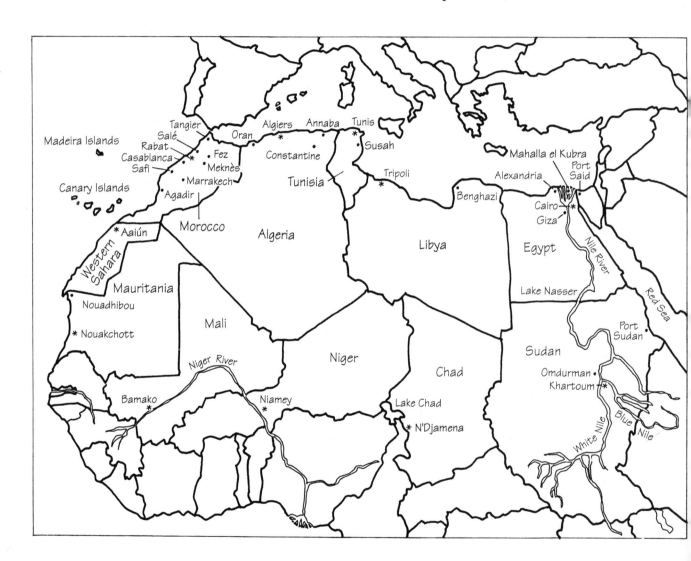

MOROCCO

The Kingdom of Morocco (or Al-Mamlakah al-Marghribiyah) is characterized by the Grand Atlas Mountains stretching from southwestern to northeastern Morocco. A wide fertile plain runs along the Atlantic coast, and a semiarid plain lies in the southeast just beyond the Atlas Mountains. The numerous rivers in the Atlantic plain are short, and the many streams on the southeastern slopes of the Atlas disappear in the Sahara.

The name Morocco originally belonged to a city in the western end of the Grand Atlas, founded during the eleventh century. Now called Marrakech (or Marrakesh), it is one of Morocco's traditional capitals and was one of the great Islamic cities during the medieval period.

TUNISIA

Tunisia, or Tunis, takes it name from the former Barbary state that once occupied the northern region of the country. A plateau region dominates western Tunisia, and a low coastal area runs along the northern and eastern coasts. A strip of the Sahara extends along the southern boundary.

West Africa

This area comprises the cluster of countries that made up colonial French West Africa, as well as the region's few former British colonies. The geography ranges from desert to tropical rain forest.

BENIN

The northern region of the Republic of Benin consists of hills and plains. Plains are also found in the east, while a marshy coastal region characterizes the south.

At the time of the Europeans' arrival, the African kingdom of Benin largely centered in the area of modern Nigeria. Much of the former Benin kingdom's territory now lies within Nigeria's boundaries. The Republic of Benin consists of territory that actually made up another African kingdom, Dahomey. Situated along what came to be known as the Slave Coast, the kingdoms of Dahomey and Benin became influential during European colonialism, largely because of their involvement in the slave trade. When Benin gained its independence in 1960, the French colony took the official name of Dahomey, only to change it to Benin in 1976.

BURKINA FASO

Savannah characterizes Burkina Faso, formerly known as Upper Volta. The northern region is grassy, while the south is sparsely forested. The former French protectorate was part of France's Upper Senegal–Niger colony until 1919, when it was made a separate colony. Achieving full independence in 1960, the republic changed its official name from Upper Volta to Burkina Faso in 1984.

CHAD

The Sahara Desert and the Tibesti Mountains traverse the northern region of Chad, one of Africa's 14 landlocked countries. Southern Chad is forested and supports agriculture. Lake Chad lies at the junction of the boundaries of Chad, Niger, and Nigeria; its southern part forms the northern extension of Cameroon. Chad was included in French Equatorial Africa from 1910 to 1920, when it became a separate French colony. It gained its independence in 1960.

GAMBIA

Gambia, or The Gambia or, in full, the Republic of The Gambia, consists of a strip of land extending for about six miles on either side of the Gambia River and about 200 miles inland from the river's mouth at the Atlantic Ocean. Bordered by Senegal on three sides, Gambia entered into a confederation with Senegal in 1982 that lasted until 1989.

GHANA

Formerly called the Gold Coast, Ghana is generally a flat area traversed by the Volta River. The country is divided between grassland plains in the north and tropical rain forest in the south.

Ghana consists of the former Gold Coast and eastern Togoland (see Togo), which were united in 1956. The territory gained its independence from Great Britain in 1957 and assumed the

name Ghana, taken from the name of an ancient Sudanese kingdom in western Africa.

GUINEA

Formerly called French Guinea, the Republic of Guinea rises from a marshy coastal plain to hilly and plateau regions, forming a tableland that provides the sources of the Niger and Senegal rivers. In southern Guinea mountain ranges reach 3,500 feet near the coast and 6,000 feet on the Liberia border.

France negotiated Guinea's boundaries with Great Britain and Portugal in the 1880s. From 1895 to 1958, Guinea was a part of French West Africa, the federation of French possessions that was officially abolished in 1959. Gaining its independence in 1958, Guinea took its name from an ancient African kingdom.

GUINEA-BISSAU

Known as Portuguese Guinea until it achieved its independence in 1974, Guinea-Bissau consists mostly of low, marshy terrain.

IVORY COAST (Côte d'Ivoire)

Its coast bordered with lagoons, Côte d'Ivoire is watered by the Bandama and Sassandra rivers and tributaries of the Niger and Volta rivers. The country slopes from the coastal plain to plateau in the central region to hills in the west and northwest. As a French colony, Ivory Coast was included in the French West Africa federation, and from 1933 to 1947 its territory incorporated Upper Volta (now Burkina Faso). The country's name officially changed to Côte d'Ivoire in 1985.

LIBERIA

A plateau country, the Republic of Liberia is densely forested and well watered, with lagoons marking its coastline. Former American slaves began settling at Monrovia in 1822, and the Free and Independent Republic of Liberia was established in 1847. Today the country includes territory that once belonged to the Republic of Maryland, founded in 1833 and united with Liberia in 1857.

MAURITANIA

The extreme western Sahara Desert covers most of the Islamic Republic of Mauritania, while mountainous plateaus rise to about 1,500 feet in the country's northern and central regions. The ancient country of Mauritania (or Mauretania) once included modern Morocco and part of Algeria, until the Romans divided it into two separate provinces. During the colonization of Africa, European nations disputed the coastal region until the 1817 Senegal Treaty recognized it as French territory. France, however, did not occupy Mauritania until the turn of the century. Constituted as part of the French West Africa federation in 1904, Mauritania became an autonomous republic within the French community in 1958 and gained full independence in 1960.

NIGER

While the Sahara Desert covers northern and central Niger, savannah characterizes the southern region. The Niger River cuts through the southwestern tip of the landlocked republic. A former native kingdom until conquered by the Berbers, Aïr (or Asben) is the mountainous region in north-central Niger.

Like Mauritania, Niger joined the French West Africa federation in 1904, became an autonomous republic within the French community in 1958, and gained full independence in 1960. From 1933 to 1947, its territory included part of Upper Volta (now Burkina Faso).

NIGERIA

Mangrove swamps are plentiful along the coastal plain of the Federal Republic of Nigeria, while the terrain is semidesert in the extreme north. The Jos Plateau lies in central Nigeria, and mountains rise in the east. The country's rivers include the Benue, the Niger River's only large tributary. Also known as Joliba, Kworra, and other native names, the Niger flows through western Nigeria to the Gulf of Guinea, where it forms an extensive delta.

SENEGAL

The Senegal River and its chief tributary, the Faleme, form the northern and eastern boundaries of Senegal. Except for a mountainous region in the southeast, the interior of Senegal is only slightly more elevated than its coast. Becoming a republic within the French community in 1958, Senegal joined the Sudanese Republic (now Mali) to form the Mali Federation in 1959 and achieved full independence in 1960. Senegal and Gambia formed the short-lived confederation of Senegambia between 1982 and 1989.

SIERRA LEONE

Mangrove swamps characterize the coastal belt of Sierra Leone where, in the late eighteenth century, English philanthropists sponsored settlements for freed and runaway slaves. Now the capital, Freetown was established on land purchased from a Temne chief in 1788. Although the settlement failed at first, it

was reestablished six years later. A British colony since 1808, Sierra Leone became independent in 1961.

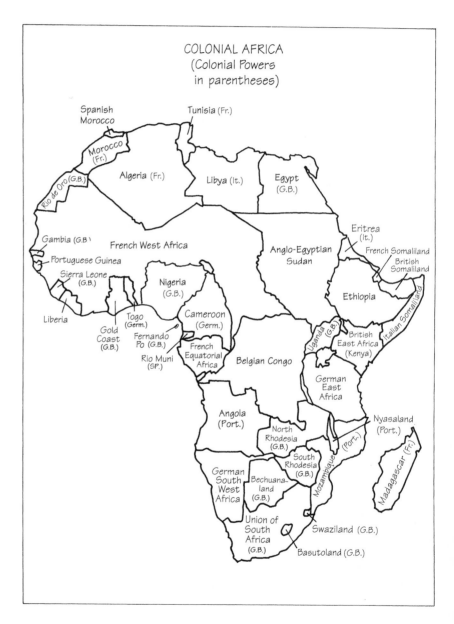

COLONIAL AFRICA
(Colonial Powers
in parentheses)

Spanish Morocco

Tunisia (Fr.)

Morocco (Fr.)

Rio de Oro (G.B.)

Algeria (Fr.)

Libya (It.)

Egypt (G.B.)

Eritrea (It.)

French Somaliland

British Somaliland

Gambia (G.B.)

French West Africa

Anglo-Egyptian Sudan

Portuguese Guinea

Sierra Leone (G.B.)

Nigeria (G.B.)

Ethiopia

Liberia

Togo (Germ.)

Cameroon (Germ.)

Gold Coast (G.B.)

Fernando Po (G.B.)

Rio Muni (SP.)

French Equatorial Africa

Belgian Congo

Uganda (G.B.)

British East Africa (Kenya)

Italian Somaliland

German East Africa

Angola (Port.)

North Rhodesia (G.B.)

Nyasaland (Port.)

South Rhodesia (G.B.)

Mozambique (Port.)

Madagascar (Fr.)

German South West Africa

Bechuanaland (G.B.)

Union of South Africa (G.B.)

Swaziland (G.B.)

Basutoland (G.B.)

TOGO

Formerly French Togo, the Republic of Togo consists of a strip of land about 70 miles wide. Its swampy coastal plain and its northern savannah are separated by a mountain range that reaches a maximum height of roughly 3,900 feet. During World War I, British and French forces captured the German protectorate of Togoland and divided it into two administrative zones. The British zone, western Togo (later called Trans-Volta Togoland), was placed under the control of Gold Coast (now Ghana) and today makes up the extreme eastern region of Ghana. The French zone became the independent Republic of Togo in 1960.

WESTERN SAHARA

Since 1979, Morocco has claimed and occupied Western Sahara, formerly called Spanish Sahara. The area to the southwest of Morocco and along the Atlantic Ocean was a Spanish outpost until 1975, when Spain divided it between Morocco and Mauritania. After Mauritania gave up its claim, Morocco occupied the entire region, which Algeria still disputes.

East Africa

The European powers once divided eastern Africa into various colonies: Belgian East Africa, British East Africa, German East Africa, and Italian East Africa (Ethiopia and part of Somalia). The Great Rift Valley (see page 103) runs through the center of this region.

BURUNDI

Known as Urundi before becoming independent in 1962, the landlocked Republic of Burundi made up the southern half of Ruanda-Urundi (see Rwanda, as it is now spelled).

DJIBOUTI

The former French Territory of the Afars and Issas, Djibouti is smaller now than the original colony, which was known as Somaliland, because France ceded some 300 square miles of the colony to Eritrea, then a part of Italian East Africa and today a separate, independent nation. The territory became an independent republic in 1977.

ETHIOPIA

Ethiopia is mountainous in its northern, southern, and central regions, with lowland on its eastern border. The Danakil Desert lies in the northeast; in the southeast the Haud Desert extends into Somalia. The country's rivers include the Takkaze and the Abay (or Blue Nile), both tributaries of the Nile River. The Awash River rises southeast of Addis Ababa, the capital, and flows northward, disappearing in the Danakil Desert.

An ancient country formerly known as Abyssinia, Ethiopia once consisted of southern

Egypt, eastern Sudan, and the northern region of modern Ethiopia, but in some classical writings, the name referred to the region extending south of Egypt to modern Tanzania. Its rulers included Menelik, said to be the son of the Hebrew King Solomon and the Queen of Sheba.

In the late nineteenth century Italy first claimed the coastal city of Aseb as a protectorate. Italy expanded its interests and eventually organized Ethiopia, Eritrea, and Italian Somaliland as Italian East Africa. The country regained its independence in 1941.

KENYA

The Great Rift Valley separates the two north-south mountain ranges in western Kenya. The republic's low coastal strip gradually extends into a wide plain, which is high and arid in the north. Leasing a coastal strip of land from the ruler of Zanzibar (see Tanzania) in 1887, the British East Africa Company extended its holdings into the interior. In 1895 the region was organized as British East Africa Protectorate and, in 1920, became a British colony, except for the coastal strip and its islands, which was named Kenya Protectorate.

RWANDA

Until gaining its independence in 1962, Rwanda made up the northern district of Ruanda-Urundi, also known as Belgian East Africa. Following World War I, the League of Nations mandated that Germany cede the German East Africa territory to Belgium. Administratively united with the Belgian Congo from 1925 to 1960, Ruandi-Urundi became two independent countries — Rwanda and Burundi — in 1962. A bloody civil war broke out in 1994, with most civilians being killed or seeking refuge across the Burundi or Tanzanian border.

SOMALIA

Most of Somalia, or officially the Somali Democratic Republic, is flat semidesert country in the central and southern regions and hilly in the north. In the Middle Ages, the Gulf of Aden region had been a powerful Arab sultanate, which was broken up in the seventeenth century. The region came under British influence in the nineteenth century, although it remained under Egyptian control until late in the 1880s. Meanwhile, the ruler of Zanzibar had leased the Indian Ocean coast to the Italians. The two areas became known as British Somaliland and Italian Somaliland until they were united to form the independent Republic of Somalia in 1960. (French Somaliland, a small area northwest of British Somaliland, became the independent Republic of Djibouti.)

SUDAN

The largest country in Africa, the Republic of Sudan is predominantly desert or semidesert, including the Nubian Desert in the northeast, where the Nubian empire once reigned. The extreme south is tropical rain forest; the central region, a grassy plain; and western and northeastern Sudan, hilly. The Nile River flows the entire length of the country.

Although Sudan was not a part of the East Africa colonies, it was under Egyptian control for most of the nineteenth century, until the Sudanese revolted in the 1880s. Egyptian and British forces again moved in and jointly administered the country until Sudan regained its independence in 1956.

TANZANIA

The United Republic of Tanzania consists of the former British colony of Tanganyika and the British protectorate of Zanzibar. Africa's highest points are found on mainland Tanzania, with the peaks of Mt. Kilimanjaro on the Kenya border rising more than 19,300 feet. Numerous rivers flow through the country; marshy regions are abundant in western Tanzania. Lakes include the southern half of Lake Victoria, about a third of Lake Tanganyika, and part of Lake Nyasa.

UGANDA

Mountains flank the Republic of Uganda on both the east and the west. Western Uganda is also noted for its dense tropical rain forest. Marshes grace the shores of Lake Victoria and Lake Kyoga. The Nile River issues from Lake Victoria, flows through Lakes Kyoga and Albert, and then northward into Sudan.

Soon after the Uganda kingdom was discovered by European explorers in 1862, religious strife combined with political rivalry to weaken the kingdom and allow the British to declare it a British protectorate in 1895. Independence was granted in 1962.

THE GREAT RIFT VALLEY

The East African Rift Valley makes up much of the Great Rift Valley, which stretches along East Africa and crosses the Red Sea into the Middle East region of western Asia. In an area where the tectonic plates are

moving apart, faults and volcanoes mark the Great Rift Valley.

The East African Rift Valley extends from the southeast coast of Mozambique northward, mostly through Malawi, Tanzania, Kenya, and Ethiopia. It includes a chain of lakes in Ethiopia; Lake Rudolf and other smaller lakes in Kenya and Tanzania; and Lakes Albert, Edward, Kivu, Tanganyika, and Nyasa. It also includes the Ethiopian Highlands (highest peak: Ras Dashen, about 15,160 feet) and the Ruwenzori Mountains, which include Mt. Kenya (17,060 feet) in Kenya and Tanzania's Mt. Meru (14,960 feet). Mt. Kilimanjaro (19,340 feet) in Tanzania, is Africa's highest peak. The Great Rift Valley extends into Asia and also includes the Sea of Galilee, Dead Sea, the Gulf of 'Aqaba, and the Red Sea, which separates Asia and Africa. The Great Rift Valley falls below sea level at the Dead Sea but rises to more than 6,000 feet above sea level in some areas of Africa.

Equatorial Africa

French Equatorial Africa, the former federation of French possessions in west-central Africa, made up a large portion of Equatorial Africa. The federation was centered around the equator and also includes former Zaire. Second in size only to the rain forest of the Amazon River basin, Africa's tropical rain forest is the region's distinguishing geographical feature.

CAMEROON

The United Republic of Cameroon is characterized by an interior plateau, with marshes along the coast and along the lower courses of rivers.

The territory comprising the republic was once part of the Cameroons, a former German colony. Following Germany's defeat in World War I, the Cameroons were divided into French and British administrative zones, which became United Nations' trust territories in 1946. The French territory became the independent Republic of Cameroon in 1960. In 1961 the southern part of the British trust territory united with Cameroon; the northern part merged with Nigeria.

CENTRAL AFRICAN REPUBLIC

The landlocked Central African Republic (or, from 1976 to 1979, Central African Empire) consists of a plateau, with savannah—drained by tributaries of the Shari River—covering the northern half of the country and tropical rain forest the southern half.

The region that would become the Central African Republic was united with Chad in 1906 to form the French colony of Ubangi-Shari-Chad. In 1910 the colony joined French Equatorial Africa. It was separated from Chad in 1920 and gained its independence in 1960.

CONGO

The name Congo derives from the Kongo, a native kingdom that occupied the northern region of modern Angola. As a region in colonial Africa, the Congo was an indefinite term for Central Africa on both sides of the Congo River. Today the Congo River and its tributary, the Ubangi, provide a large portion of the boundary between the countries of former Zaire and Congo.

Formerly called Middle Congo, Congo became a part of French Equatorial Africa in 1910. It gained independence in 1960, and its official name became the Republic of the Congo.

EQUATORIAL GUINEA

Formerly Spanish Guinea, Equatorial Guinea consists of Rio Muni province on mainland Africa, the Fernando Po Islands west of Cameroon, and Annobon Island west of Gabon, as well as several other smaller islands just off the southwestern tip of Equatorial Guinea. The capital, Malabo, is located on Fernando Po.

With its volcanic formations, Fernando Po has rich soil and an average annual rainfall of 100 inches. Beyond its narrow coastal plain, Rio Muni is a series of plateaus. The Spanish territory gained full independence in 1968.

GABON

The basin of the Ogooué River covers most of Gabon, known officially as the Gabonese Republic. North and south of the basin the country is hilly.

DEMOCRATIC REPUBLIC OF THE CONGO

The territory occupying the greater part of the Congo River basin (formerly called Zaire) began as Congo Free State, established in 1885 by the International Association of the Congo, a private organization controlled by the king of Belgium, Leopold II. Belgium annexed the territory in 1908, renaming it Belgian Congo.

Officially the Democratic Republic of the Congo since its independence in 1960, it claims numerous tributaries of the Congo River. Most of the country consists of a low plateau, with marshes along the Congo in the northwest. The country's lakes include Tanganyika, Mweru, Kivu, Edward, and Albert, all of which the Democratic Republic of the Congo shares with its eastern neighbors.

Southern Africa

The southern region of the continent includes former Portuguese colonial possessions as well as the Dutch legacy in the Republic of South Africa. The Kalahari and the Namib deserts are found in the region.

ANGOLA

The interior of Angola, formerly Portuguese West Africa, forms a part of the Central African Plateau, a roughly 100-mile coastal plain that varies from 30 to 100 miles wide. And among Angola's rivers, the Congo River forms about 100 miles of the country's border with the Democratic Republic of the Congo. Angola's territory includes Cabinda, a small district separated from Angola by the Democratic Republic of the Congo's 25-mile coastline.

The Portuguese settled São Paulo de Luanda (now called Luanda, the capital) in 1575. Independence was granted in 1975.

BOTSWANA

The great Kalahari Desert spreads across southwestern and western Botswana, formerly Bechuanaland. Salt lakes and the Okovango River basin characterize the north. The country receives an average of only 18 inches of rain per year.

In 1885, Great Britain divided Bechuanaland, attaching the southern region, British Bechuanaland, to the Cape of Good Hope (see South Africa, page 108). North of the Molopo River, Bechuanaland remained a protectorate until it became an independent republic in 1966.

LESOTHO

The mountainous Kingdom of Lesotho (formerly named Basutoland) lies completely within the Republic of South Africa. The Caledon River is Lesotho's border along the west. The Orange River flows south and southwest through the country. The British annexed Basutoland, making it a part of its Cape Colony, in 1871, and a separate colony from 1884 until its independence in 1966.

MALAWI

The Great Rift Valley traverses Malawi (formerly Nyasaland). Lake Nyasa extends for 360 miles through eastern Malawi, with its outlet, the Shire River, crossing the southern region. A British protectorate from 1953 to 1963, Nyasaland was included in the Federation of Rhodesia and Nyasaland.

MOZAMBIQUE

The Zambezi River cuts through central Mozambique and flows into the Mozambique Channel. Other rivers include the Rovuma River, which marks the country's northern boundary. Mountainous in the north and along

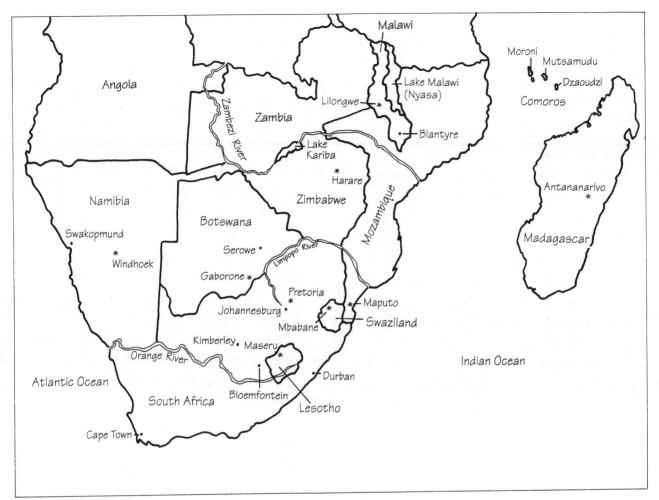

the Zimbabwe border, Mozambique also has an extensive coastal plain.

The town of Mozambique was a Portuguese trading fort early in the sixteenth century, but the territory, known as Portuguese East Africa, wasn't organized into a colony until 1907. Granted independence in 1975, the new nation took the name of its original Portuguese town.

NAMIBIA

Once known as South-West Africa and formerly as German Southwest Africa, Namibia is largely plateau, with the Namib Desert extending along the coast.

The Union (now Republic) of South Africa captured German Southwest Africa during World War I and gained control of the colony as a mandate from the League of Nations in 1919. In 1966, South Africa rejected a United Nations resolution declaring the mandate terminated, and did not recognize the United Nations name of Namibia for the territory. In 1990 Namibia became an independent country.

SOUTH AFRICA

The physical features of the Republic of South Africa range from grassland (or *veld,* a Dutch word meaning "field") in the northeast and between the Great Karoo Mountains and the Orange River to desert or semidesert in much of the west and south, with the southern part of the Kalahari Desert lying between the Orange and Molopo rivers. The Orange River, with its tributary the Vaal, traverses central South Africa, flowing westward to the Atlantic Ocean. The heart of South Africa, however, is a plateau rimmed by the Great Escarpment, a sometimes sheer and often thousands-of-feet-high barrier separating the narrow coastal plain from the extensive veld.

The southern tip of Africa has the continent's greatest ethnic complexity. Long before the arrival of Europeans, the population of southern Africa included the Bantu, which comprise four main ethnic groups and include the Zulu and Xhosa people; and the Khoisan (later termed Hottentots and Bushmen by the whites), who are almost nonexistent today.

Although the Cape of Good Hope was discov-

ered in 1488, it wasn't until the mid-seventeenth century that the Dutch founded the settlement of Cape Town, which served as a stopover for ships en route from Europe to India. In the early nineteenth century, the Dutch ceded the province of Cape Hope to the British. The Boers, descendants of the original Dutch settlers and later known as Afrikaaners, trekked north and east of the Orange River, eventually forming the South African Republic (now the province of Transvaal) and Orange Free State.

Following the Boer War, negotiations between the Boers and the British led to the formation, in 1910, of the Union of South Africa, with three capitals: the administrative capital in Pretoria, also the capital of Transvaal; the judicial capital at Bloemfontein, the capital of Orange Free State; and the legislative capital at Cape Town, the capital of Cape Province. Growing in political strength, the Boers were able to win control of the government and in 1961 proclaimed the Republic of South Africa and withdrew from the (formerly British) Commonwealth of Nations.

The political geography of South Africa had been dominated by the system of apartheid from 1948-1994. In Afrikaans (the language developed from Dutch and indigenous dialects), the word *apartheid* means "apartness" or "separation." The objective being the separation of ethnic groups and cultural communities, apartheid was not only to separate whites from people of color but, within these groups of colored peoples, to separate the blacks from Asians, the Asians from Zulus, and so on. English-speaking whites were viewed separately from Afrikaans-speaking whites. The grand design was a South Africa with racially based "homelands." The homelands, however, were impoverished and their black inhabitants had to apply for temporary work permits outside the homelands in factories, mines, and elsewhere in white South Africa.

In 1994, a negotiated agreement between the African National Congress and the administration of president F.W. de Clerk abolished apartheid, and democratic elections were held for the first time. Blacks and other non-whites were given equal rights under the presidency of Nelson Mandela and his multi-party government.

SWAZILAND

Settled by the Swazi branch of the African Zulu nation in the early 1880s, the Kingdom of Swaziland consists of grassland. Although the British had guaranteed the kingdom's independence, they took over administrative control after the Boer War at the turn of the century. In 1968, Swaziland gained independence.

ZAMBIA

Formerly Northern Rhodesia, Zambia consists of tableland crossed by three main rivers: the Kafue, the Luangwa, and the Zambezi, which forms the boundary with Zimbabwe. Zambia and Zimbabwe share the Zambezi River's Victoria Falls.

Made a British protectorate (one level below colony classification), Northern Rhodesia was part of the Federation of Rhodesia and Nyasaland (see Zimbabwe) from 1953 to 1963. It became the independent Republic of Zambia in 1964.

ZIMBABWE

Known as Rhodesia or Southern Rhodesia before 1979, and Zimbabwe-Rhodesia from 1979 to 1980, Zimbabwe forms part of the South Africa plateau. Central Zimbabwe is a broad watershed between the Zambezi (on the boundary with Zambia) and the Sabi and Limpopo river systems.

Made a self-governing British colony in 1923, the colony was part of the Federation of Rhodesia and Nyasaland.

A Continent of Extremes

The continent is nearly bisected by the equator, with the climatic zones of northern and southern Africa mirroring each other from the tropical rain forests in equatorial Africa through the semitropical savannahs, dry steppes, deserts, and Mediterranean coastal climates. Tropical or subtropical climates characterize two-thirds of Africa and make it the hottest continent on Earth. While the forests may receive heavy rains, other parts get less than 10 inches of precipita-

Water Facts

Africa is crisscrossed with rivers and dotted with lakes and waterfalls.

Rivers

The Nile begins in south-central Burundi and flows northward before it empties into the Mediterranean Sea. It is over 4,000 miles long.

The Zaire (still known as the Congo in Congo) begins at the confluence of the Lualaba and Luapula rivers in former Zaire, flows northward in a great arc, and empties into the South Atlantic Ocean. It is over 2,700 miles long.

The Niger begins in Guinea near Sierra Leone. It flows northeast through Guinea and Mali, then southwest through Niger and Nigeria on its way to the Gulf of Guinea. It is over 2,600 miles long.

The Zambezi starts in northwest Zambia and follows an S-shaped route through Angola, Zambia, and Mozambique before emptying into the Indian Ocean. It is over 1,700 miles long.

The Orange begins in the Drakensberg Mountains in Lesotho and forms the border between Lesotho and Orange Free State in South Africa. It flows through Namibia, before emptying into the Atlantic Ocean. It is over 1,300 miles long.

The Limpopo begins north of Transvaal Province, South Africa, and travels along the border with Botswana and Zimbabwe before emptying into the Indian Ocean. It is 1,100 miles long.

The Ubangi begins at the confluence of the Uele and Bomu rivers in former Zaire. It forms part of the border between former Zaire and the Central African Republic on its way to the Zaire River. It is approximately 700 miles in length.

Lakes

Lake Victoria occupies roughly 26,900 square miles in southern Uganda and is Africa's largest lake. It is the third largest lake in the world.

Lake Tanganyika in western Tanzania and eastern Democratic Republic of the Congo is 12,700 square miles and has a maximum depth of almost 5,000 feet.

Lake Malawi in eastern Malawi occupies more than 11,000 square miles and is over 2,280 feet deep in places.

Lake Chad occupies over 6,300 square miles.

Lake Volta in Ghana is over 3,000 square miles.

Lake Rudolf (Turkana) in northwestern Kenya is almost 2,500 miles in area.

Lake Albert in northeastern Democratic Republic of the Congo and northwestern Uganda is over 2,000 feet in depth.

Lake Kariba in Zambia and northwestern Zimbabwe is over 2,000 square miles in area.

Waterfalls

Tugela in Natal, South Africa, is over 3,000 feet high.

Kalambo in Tanzania and Zambia: 720 feet.

Maletsunyane in Lesotho: 630 feet.

King George's in South Africa: 400 feet.

Victoria in Zimbabwe and Zambia: 355 feet.

tion a year. Devastating droughts are common, among the most recent occurring in Ethiopia.

Second in size only to the Amazon River basin, a great expanse of tropical rain forest extends across much of central Africa. It extends across the West African coastal region from Gambia to Cameroon and also covers Gabon, Congo, and half of the Democratic Republic of the Congo.

Africa's Deserts

Not only does Africa contain the world's longest river (the Nile) and the second largest freshwater lake (Victoria), it also lays claim to the world's largest desert. With an average width of nearly 900 miles, and ranging from 100 feet below sea level to more than 11,000 feet above sea level, the Sahara covers roughly 3.5 million square miles of North Africa, or nearly a third of the continent. The desert extends from the Red Sea to the Atlantic Ocean and covers all, most, or part of 11 countries—Egypt, Libya, Tunisia, Algeria, Morocco, Western Sahara, Mauritania, Mali, Niger, Chad, and Sudan—and includes the Nubian (in Sudan), Libyan (in eastern Libya and southwestern Egypt), and Eastern Arabian (in eastern Egypt) deserts. The Sahara has areas of drift sand and dunes, rock, gravel, and pebbles.

Roughly 100,000 square miles between the Orange and Zambezi rivers, southern Africa's Kalahari Desert region is largely savannah, subtropical grasslands with scattered woodland, characterized by drought. At an average elevation of 3,000 feet, the Kalahari extends across southwestern Botswana, southeastern Namibia, and northern Cape Province, South Africa. In the extreme southwestern reaches of the Kalahari, the subtropical geography turns to more desertlike conditions. Although the overall rainfall average is about 8 inches a year, the annual rainfall ranges from 6 inches or less to as much as 20 inches a year.

The Namib Desert lies in western Namibia, along most of the country's Atlantic coastal region. Covering some 800 square miles, the desert contains the highest sand dunes in the world.

Quiz

(answers, p. 113)

1. In what country would you be if you were viewing the pyramids?

2. In their official anthem, the U.S. Marines proclaim, "From the halls of Montezuma to the shores of Tripoli." The halls of Montezuma refer to Mexico. Where is Tripoli?

3. What African country, founded by freed American slaves in 1817, named its capital after the fifth president of the United States?

4. What six African countries are bisected by the equator?

5. What Egyptian city was founded by Alexander the Great in 332 BC?

6. In what country does the Humphrey Bogart film *Casablanca* take place?

7. Following his final defeat at Waterloo in 1815, French emperor Napoleon Bonaparte abdicated and surrendered to the British, who exiled him to St. Helena. Where is St. Helena?

8. Where do Tangerines live?

9. Made of red felt and usually adorned with a tassel, the fez is most often identified with eastern Mediterranean countries. But the brimless, cone-shaped, flat-crowned hat was actually named for the city where it was first made. In what country is the city located?

10. What country is famous for its diamonds?

11. The film *Gorillas in the Mist* starred Sigourney Weaver as primatologist Dian Fossey, who spent much of her life studying the ancient mountain gorilla. In what Central African country are you most likely to find the mountain gorilla?

12. Danish baroness Karen Blixen became a famous novelist under the pseudonym Isak Dinesen. In what East Africa country did she live?

13. In 1976, Entebbe was the scene of the Israeli forces' dramatic rescue of a plane that had been hijacked by Palestinian terrorists. Where is Entebbe?

14. What modern country's name sprang from the unification of Tanganyika and Zanzibar?

15. Where is the Barbary Coast, famous for piracy from the sixteenth century to the eighteenth century?

16. In what modern country was the ancient city of Carthage?

17. In what country would you find people speaking Afrikaans?

18. Identify the modern countries formerly known by the following European names:
    a. Rhodesia (Southern Rhodesia)
    b. Northern Rhodesia
    c. Belgian Congo
    d. Portuguese West Africa
    e. Abyssinia
    f. South-West Africa

19. In May 1967 secessionists set up

the Republic of Biafra, precipitating a civil war that ended in January 1970 with Biafra's surrender to the federal government. The area is part of what country?

ANSWERS

1. Egypt

2. Originally a Phoenician colony, Tripoli (Greek for "three cities") was named for its three chief cities: Oea, Leptis Magna, and Sabrata. Today Tripoli is the capital of Libya.

3. Monrovia, the capital of Liberia, was named after James Monroe, U.S. President from 1817 to 1825.

4. Gabon, Congo, Democratic Republic of the Congo, Uganda, Kenya, and Somalia

5. Alexandria

6. Casablanca is in Morocco.

7. A British territory, St. Helena is a volcanic island in the South Atlantic Ocean.

8. Tangiers, Morocco

9. Morocco

10. South Africa

11. Rwanda

12. Kenya

13. Entebbe is the former capital of Uganda.

14. Tanzania

15. The coast of North Africa. The Barbary Coast and the Barbary states of Morocco, Algeria, Tunisia, and Tripolitania (northern Tunisia and Libya) take their names from Barbarossa, who led the Turkish conquest of the area in the 1530s.

16. Tunisia

17. Largely Dutch in origin, Afrikaans, also known as South African or Cape Dutch, developed in South Africa among the descendants of Dutch settlers. It also incorporates words from Bantu and other native African languages spoken in southern Africa long before the arrival of the Dutch settlers. There is also a pidgin Afrikaans used for communication between black Africans and Afrikaaners.

18. a. Zimbabwe
    b. Zambia
    c. Democratic Republic of the Congo
    d. Angola
    e. Ethiopia
    f. Namibia

19. Nigeria

Arctic Ocean

East Siberian
Sea

Bering Sea

Kara Sea

Laptev
Sea

Russia

Sea of
Okhotsk

Black Sea

Turkey

Cyprus

Israel

Lebanon

Syria

Caspian Sea

Kazakhstan

Uzbekistan

Mongolia

North
Korea

Sea
of
Japan

Japan

Jordan

Iraq

Turkmenistan

Kyrgyzstan

Iran

Tajikistan

Red Sea

Kuwait

Persian
Gulf

Afghanistan

China

Yellow
Sea

South
Korea

Bahrain

Qatar

Pakistan

Nepal

East
China
Sea

Pacific
Ocean

Saudi Arabia

Bhutan

India

Yemen

Oman

United Arab
Emirates

Vietnam

Taiwan

Gulf of Aden

Bangladesh

Hong Kong

Arabian Sea

Myanmar

Laos

Macao

Philippines

Bay of Bengal

Thailand

South
China
Sea

Sri
Lanka

Cambodia

Brunei

Malaysia

Maldives

Singapore

Indonesia

Indian Ocean

114

ASIA

The world's largest continent, Asia incorporates about one-third of the world's landmass and more than half of the world's population. It includes some of the oldest and largest cities on earth—Beijing, Shanghai, Yokohama, Tokyo, Jerusalem, and Baghdad. Its more than 17.1 million square miles range from the frozen plains of Siberia in the north to the tropical rain forests of Malaysia in the south, from the Western Arabian Desert to the eastern volcanoes of Japan.

Asia is bounded on the north by the Arctic Ocean and its subdivisions, including the Kara, Laptev, and East Siberian seas; on the south by the Indian Ocean and its subdivisions (the Bay of Bengal, the Arabian Sea, the Gulf of Aden, and the Persian Gulf); on the east by the Pacific Ocean (Bering Sea, Sea of Okhotsk, Sea of Japan, and the Yellow, East China, and South China seas); and on the west by the Ural and Caucasus mountains, the eastern banks of the Volga River, and the northwestern banks of the Caspian Sea (the conventional boundary between Asia and Europe), as well as the isthmus of Suez and the Red, Mediterranean, Aegean, and Black seas.

Excluding the Commonwealth of Independent States (see p.80), Asia has four regions: eastern Asia, Southeast Asia, southern Asia, and western Asia.

Eastern Asia

Often referred to as the Far East, eastern Asia includes China, Hong Kong, Japan, Korea (North and South), Macao, Mongolia, and Taiwan, with thousands of islands off the coasts of Japan and Taiwan forming the eastern boundary. On its western boundary, the Pamir Mountains separate China from India and Afghanistan, with the high plateau of Tibet encircled by two great mountain chains, the Himalayas and the Kunlun Shan, which extend into China.

While Tokyo alone may receive as much as 48 inches of rain a year, mostly from May to September, less than one inch of rain each year falls in some parts of both China and Mongolia. This sparse precipitation has created the Gobi and

Talka Makan deserts. Yet the coastal strip—the plains and valleys of Taiwan, Korea, and China—are wet enough to produce abundant rice crops.

CHINA

China's three major rivers—the Yellow, the Yangtze, and the Hsi—have been great commercial highways for centuries. The Himalayas run along the south and southwest border of Tibet, itself a great plateau with an elevation of more than 10,000 feet. Other mountain ranges include the Tien Shan in western Xinjiang Uygur and the Great Khingan range in Inner Mongolia. China has many shorter and lower ranges, especially in the south and west. The highest known peak is Gongga Shan (24,900 feet) in Sichuan. Mostly in the Mongolian People's Republic and Inner Mongolia, the Gobi Desert extends for 500,000 square miles; the southwestern portion is entirely sand, but elsewhere is steppe land.

Chinese civilization probably spread from the Yellow River valley, where it existed around 3000 BC. The Chou dynasty was the first known and extended from the Yangtze River to the Great Wall in the north. The Chinese border continued to expand and once included northern Korea. In the thirteenth century all of China was included in the Mongol empire, which stretched across Asia into Europe as far as Lithuania. (Visited by Marco Polo in the late thirteenth century, the Mongols' Yuan dynasty was ruled by Kublai Khan.)

The Chinese empire at various times included Turkestan (now part of the Commonwealth of Independent States), Korea, Annam (today a part of Vietnam), Siam (Thailand), Burma (Myanmar), and Nepal. Its 1689 treaty with Russia defined the two countries' border. China ceded eastern Siberia as far as Vladivostok to Russia in 1858 and lost Korea and Taiwan to Japan in 1895. Following civil war (1945–1950), the Communist regime established control over mainland China, while Chiang Kai-shek and his followers in the nationalist government fled to Taiwan. In 1997 British Hong Kong reverted to Chinese control, and in 1999 Portuguese Macao was returned to China.

JAPAN

A chain of islands off the eastern coast of Asia, Japan (or Nippon, in Japanese) comprises more than 143,600 square miles and has four main islands: Honshu, Shikoku, Kyushu, and Hokkaido. All the islands are mountainous and include many high peaks and volcanoes. The highest is Mt. Fuji, on Honshu. Japan lies within the earthquake belt of the western Pacific Ocean and

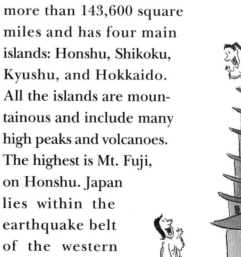

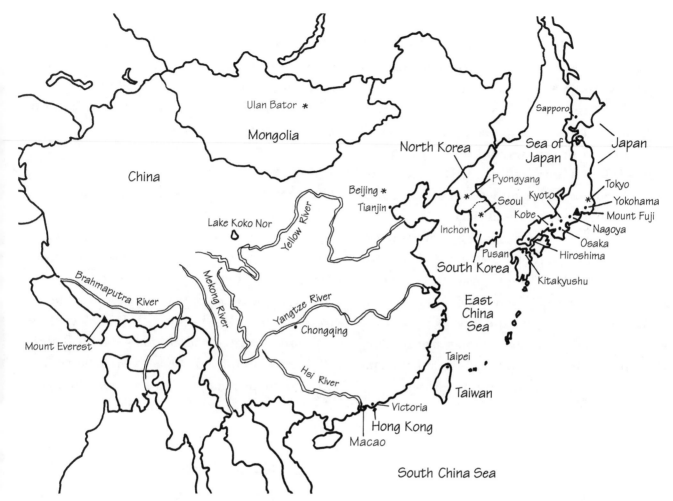

has suffered from many destructive shocks. The islands are indented with many bays, affording fine harbors, although only a few of the short rivers are even partly navigable. The islands also boast numerous lakes, including Inawashiro and Biwa.

In 794 the imperial capital was moved to Heian-kyo (modern Kyoto). In 1868 the capital was moved to Edo, a city on the island of Honshu, with the name changed to Tokyo.

In the twentieth century Japanese territory expanded beyond the island group when Japan annexed Korea. For participation in World War I, it received Germany's Pacific possessions north of the equator. But it returned to the boundaries of the island group after World War II.

KOREA

A peninsula on the east coast of Asia, the former Kingdom of Korea was partitioned into two republics—South Korea and North Korea.

North Korea is generally mountainous, especially in the northern regions; its western coast has numerous estuaries and tidal flats. South Korea is mountainous in its eastern and south-central regions and includes numerous islands off its southern and southwestern coasts.

The Kingdom of Choson was established as early as the twelfth century BC. The Chinese and Japanese have long fought over Korea. Japan annexed Korea in 1910 and held it until its defeat in World War II. After the war, as part of the spoils of the war, the kingdom was divided along the 38th parallel into two zones of occupation, the Soviet in the north and the American in the south. North Korea became the Democratic People's Republic of Korea; South Korea, the Republic of Korea.

MONGOLIA

The country usually called Mongolia is formally named Mongolian People's Republic and does not include Inner Mongolia in northwestern China nor the Autonomous Republic of Tuva to the north, which was formerly associated with the Union of Soviet Socialist Republics.

The People's Republic comprises more than 604,000 square miles and is surrounded by China

and the independent state of Russia. Western Mongolia is mountainous, while much of the central and southeastern part of the country is dominated by the Gobi Desert. Its chief rivers, the Selenga and its tributary the Orhon, flow into Lake Baikal, to the north in Russia.

TAIWAN

Off the coast of southeastern China, Taiwan and its outlying islands comprise nearly 13,900 square miles. Lofty mountains rise in east-central Taiwan and there are no long rivers. Its minor dependencies include Pescadores.

Taiwan became the seat of China's Nationalist government following the 1945–1950 civil war when the Chinese Communists established the People's Republic of China on the mainland. In 1971, when the People's Republic was admitted to the United Nations, Taiwan lost its seat in the organization, its legitimacy no longer recognized.

Southeast Asia

Southeast Asia was named by Europeans around 1900 to define a specific trading area, but the Chinese and Japanese had long used a similar term to describe the region. Southeast Asia embraces Brunei, Indonesia, Cambodia, Laos, Malaysia, Myanmar (Burma), Singapore, Thailand, Vietnam, and the Philippines in a tangle of seas and straits. It is an area of peninsulas and islands, bordered by staggering ocean

depths: the Java Trench, the deepest known part of the Indian Ocean (25,344 feet), and the Mariana Trench, with the deepest known part of the Pacific Ocean, Challenger Deep (approximately 36,200 feet) in the Mindanao Trench.

The mountains running through the Indonesian islands of Sumatra and Java are volcanic in origin. Some 70 eruptions have been recorded in the last 200 years.

The equator bisects Indonesia, creating a year-round hot and humid climate and the densest tropical rain forests in the world. The areas north and south of the islands are rich in temperate rain forest with dwarf trees and shrubs, teak and pine forests, and wooded savannahs.

Each mainland country has a major river valley: the Irrawaddy in Myanmar (central Burma), the Chao Phraya in Thailand, the Red in northern Vietnam, and the Mekong running through Laos, Cambodia, and southern Vietnam.

There is a vast array of religions practiced in this region: Buddhism in Myanmar and Thailand (and in Laos, Cambodia, and Vietnam, where Communism has attempted to obliterate the religion); Islam in Malaysia, Brunei, and Indonesia; Hinduism in Bali, where tribal gods are still worshiped in the hills; and Catholicism in the Philippines.

BRUNEI

A sultanate in northeast Borneo, Brunei is divided geographically into two parts. These regions are separated by the Limbang River valley. Both areas are surrounded by Sarawak (Malaysia) and have a coastline on the South China Sea and Brunei Bay.

CAMBODIA

Cambodia comprises nearly 70,000 square miles of generally level land, with a large jungle area and the Dangrek Mountains along the northern border. Most of Cambodia lies in the basin of the Mekong River.

In ancient times the Khmer kingdom flourished around the city of Kambodja, founded around the year 435. At its height, from the ninth to the twelfth centuries, the kingdom ruled the entire Mekong Valley and the Shan states along the Mekong's tributaries. After the thirteenth century it became a province alternately of Annam (Vietnam) and Siam (Thailand), and the borders have long been the cause of wars with Vietnam and Thailand.

INDONESIA

An archipelago surrounded mostly by the South China Sea, the Pacific Ocean, and the Indian Ocean, the Republic of Indonesia comprises nearly 780,000 square miles and includes the islands of Bali, Borneo, Java, Sumatra, and the western portion of New Guinea known as Irian Jaya.

Formerly known as the Netherlands East Indies or the Dutch East Indies, the United States of Indonesia was established in 1949, followed in 1950 by the establishment of the Republic of Indonesia.

LAOS

With more than 91,000 square miles, the thickly forested Laos is a mountainous country, with its northern peaks rising more than 9,000 feet and its southern peaks about 5,000 feet. Although it occupies the Mekong River valley in the northwest, most of the country in the north and south lies east of the Mekong River, which forms a large part of the boundary with Thailand.

MALAYSIA

The Federation of Malaysia consists of 11 states on the Malay Peninsula (West Malaysia), south of Thailand, and two states on the island of Borneo (East Malaysia), together comprising about 128,700 square miles. Singapore seceded from the federation in 1965.

MYANMAR (Burma)

The name Myanmar comes from the Burmese name for the Socialist Republic of the Union of Burma, nearly 271,800 square miles constituting the northwestern region of the Indochina Peninsula. Northern Myanmar is mountainous, while the basins of the Irrawaddy River and its tributaries occupy most of the country. Inhabited by people of Mongolian ancestry and probably of Tibetan origin, Myanmar was settled on the coast and at river mouths by Hindus in the third century. The country went through numerous divisions and reformation in the nineteenth century, following initial conflicts with the English East India Company. A former province of British India, Myanmar was separated from India and made a crown colony in 1937 until its independence in 1947. Border disputes with China were finally settled in 1960.

THE PHILIPPINES

An archipelago of some 7,100 islands, the Republic of the Philippines extends north to south for about 1,150 miles and east to west for about 690 miles and comprises 115,650 square miles. The islands are part of the western Pacific Ocean's volcanic chain, with all of them mountainous; about 20 of the mountains are volcanoes.

A Spanish colony first settled in 1565, the Philippines were ceded to the United States after the Spanish-American War in 1898. The United

States established the Commonwealth of the Philippines in 1934, with independence coming in 1946.

SINGAPORE
The 225-square-mile Republic of Singapore occupies Singapore Island, off the south coast of the Malay Peninsula, and several adjacent islets.

Following a two-year history as a member of the Federation of Malaysia, Singapore seceded and became an independent republic in 1965.

THAILAND
Formerly known as Siam, the Kingdom of Thailand comprises nearly 199,000 square miles and extends into the Malay Peninsula. North-

western Thailand is a mountainous region, while central Thailand is a plain in the basins of the Chao Phraya and its tributaries, and eastern Thailand lies in the basins of the Mekong River and its tributaries.

Part of the Mon-Khmer kingdom in ancient times, the Thai people didn't form a separate state until 1350. Beginning in the nineteenth century, Thailand began renouncing territorial claims or actually ceding territory, which included claims on Cambodia, and yielding four unfederated Malay states. Seized by Japan in 1943, it gained two Shan states and the four unfederated Malay states, which it then relinquished after Japan's defeat in World War II.

VIETNAM

Sometimes written as "Viet Nam," Vietnam comprises more than 130,400 square miles on the eastern Indochina Peninsula. The north is mountainous, while much of the south is characterized by the Mekong Delta, a flat, marshy region very suitable for the cultivation of rice. The Red River delta in the north is also an important agricultural region.

Becoming a province of China in 111 BC, Vietnam remained under continuous Chinese control for nearly 1,000 years. In the seventeenth century southern Vietnamese built a wall dividing the country into north and south, the north later known as Tonkin and the south as Cochin

China, which were then unified under a single dynasty in 1802.

Following the French defeat in 1954, after eight years of war between southern Nationalists and northern Communists, an international conference in Geneva, Switzerland, partitioned Vietnam along the 17th parallel, almost exactly along the same lines as the southern Vietnamese wall in the seventeenth century. North Vietnam (or Democratic Republic of Vietnam, with its capital at Hanoi) and South Vietnam (or Republic of Vietnam, with its capital at Saigon, or Sai Gon) were reunited in 1976, following the end of the Vietnam War. The country's official name became Socialist Republic of Vietnam, with its capital at Hanoi, and Saigon was renamed Ho Chi Minh City.

Southern Asia

Kite-shaped southern Asia extends about 1,800 miles from the Himalayas to the tip of India and about the same distance at its broadest, from Pakistan in the west, through India and the Himalayan countries of Nepal and Bhutan, to Bangladesh, including the island nation of Sri Lanka off the southeast coast of India. The region encompasses 18 principal languages with innumerable dialects, including some 500 forms of Indo-Aryan, the easternmost branch of the Indo-European languages exemplified by Punjabi, Gujarati, and Marathi.

The northern area of southern Asia is dominated by the Himalayas, and the southern area contains the Indian Peninsula, including Sri Lanka, a continental island nation off the coast of India. In between these two regions lie the Indus and Ganges rivers, the Great Indian Desert, and the Deccan Peninsula.

The area considered southern Asia is actually a subcontinent, a part of the Indian-Australian tectonic plate that has pushed against the Eurasian plate and forced up the Himalayas. The mountains basically separate Asia from the subcontinental countries of Bangladesh, Bhutan, Nepal, Pakistan, Sikkim, and, of course, the Republic of India.

If it were physically separated from the rest of Asia, the subcontinent of India would be the continent of India.

The area was being invaded by peoples from the west as early as 1500 BC, when Aryans from the Iranian Plateau pushed the Dravidians and Mundas to the south and southeast. Some of the world's most important religions were beginning on the subcontinent almost as early: Buddhism, for one, was founded in the sixth century BC; Brahmanism, with its social caste system, evolved from the Vedic religion of the Aryans. Hinduism, then Islam, and later Christianity took hold.

India Facts

India is considered a vast subdivision of Asia. It is actually the exposed part of a separate tectonic plate, the Indian-Australian plate. The end of the Indian-Australian plate is pressing against the Eurasian plate, and the pressure between the two is causing the Himalaya Mountains—already the world's highest peaks—to continue rising.

The English name *India* comes from the Sanskrit word *sindhu*, used to describe the ancient civilization in the Indus Valley. In Greek that word became *sinthos*; in Latin, *sindus*. Corrupted to *indus*, which means river, it was first applied to the region that now forms the heart of Pakistan. (Indus River, then, means "River River," or perhaps "River of Rivers.") Indus became India, referring to the land of river basins and the people from the Indus River to the Brahmaputra River in the east.

Geofact

At an elevation of more than 11,800 feet, Lhasa, Tibet, is the highest capital in Asia, and the second highest in the world (La Paz, Bolivia, is 200 feet higher).

When the British established trading posts on the subcontinent in the seventeenth century, the region was populated by numerous tribes and kingdoms of various religions. The British began acquiring territory, and by the end of the nineteenth century, the only parts of India not under British control were protected states under native rulers, which had varying degrees of independence from the British.

British rule in India ended in 1947 with the creation of the sovereign states of India and Pakistan. Within a year the two states clashed over possession of Jammu and Kashmir, two princely states now divided between India and Pakistan.

BANGLADESH

Bisected by numerous rivers and streams and having an average rainfall of 60 inches per year, much of the generally flat Bangladesh is subject to flooding. The country, comprising more than 55,000 square miles, was originally part of Bengal. A former province of India, Bengal was divided between India and Pakistan in 1947; West Bengal fell within the borders of India, while East Bengal became part of Pakistan, about 1,000 miles to the west, and was known as East Pakistan. East Pakistan declared its independence in 1971, and following its victory—with Indian military support—officially became Bangladesh.

BHUTAN

A kingdom in the eastern Himalayas, Bhutan is marked by deep valleys and mountain peaks as high as 24,000 feet above sea level. Portions of the kingdom were annexed by India in 1865. China laid territorial claims in the late 1950s.

INDIA

The nearly 1,230,000-square-mile Republic of India can be divided into three geographical regions: the Himalayan region in the north; the Indo-Ganges Plain between the river basins of the Ganges and the Indus rivers, from the foothills of the Himalayas to those of the Vindhya Mountains; and the plateau region in central and southern India, known as the Deccan Peninsula. The Ganges and its chief tributary, the Yamuna, run across northern India; the Indus and its tributaries are in the northwest.

About 15 years after India and Pakistan settled their dispute over Jammu and Kashmir, India annexed the Portuguese territories of Goa, Daman, and Diu on the west coast of India. In 1975, India gained additional territory when the Kingdom of Sikkim abolished its monarchy and became an Indian state.

MALDIVES

A group of 19 coral atolls in the Indian Ocean, the Maldives lie southwest of Sri Lanka, where the principal industry is fishing. The capital is Male.

NEPAL

The central and northern areas of the 54,000-square-mile Kingdom of Nepal are dominated by the Himalayan ranges. Nepal's—and the world's—highest peak is Mt. Everest, which it actually shares with Tibet, a formerly autonomous region of China. In the south Nepal is cultivated and forest land.

PAKISTAN

The 310,400 square miles of the Islamic Republic of Pakistan vary from desert and marshes and barren land in the southwest to the mountains in the northeast and the fertile valleys of the Indus and its tributaries in the east. In addition to the Jammu-Kashmir settlement with India, Pakistan's northwestern border with China wasn't set until the early 1960s. The republic originally consisted of two areas, known as East Pakistan and West Pakistan, separated by about 1,000 miles of Indian territory. East Pakistan became Bangladesh in 1971.

SRI LANKA

Formerly Ceylon, Sri Lanka is an independent state in the Indian Ocean, south of India. A chain of shoals separating Palk Strait and the Gulf of Mannar, Adam's Bridge connects the 25,300-square-mile island to India.

Western Asia

Before World War I western Europeans considered western Asia as two distinct entities: the Near East or Ottoman Empire (Turkey and the countries bordering the Mediterranean Sea), and the Middle East (Afghanistan and the countries around the Persian Gulf). After 1918 and the fall of the Ottoman Empire, however, the

Desert Shield

During the Persian Gulf war in early 1991, the construction of bunkers and military installations, as well as the Allied and Iraqi bombing raids, left thousands of craters in the Arabian Desert. The removal of the natural "desert shield," small rocks and pebbles that cover the sand, increased the formation of sand dunes. The desert winds blow the newly exposed sand into dunes that can grow as high as 200 feet, and the dunes creep at a rate of as much as one foot per day. The sand keeps piling higher and higher and people have to leave because they can't dig out from the sand. Sand dunes and sandstorms can engulf roads, plantations, airports, and even whole villages in Kuwait, Saudi Arabia, and southern Iraq.

Near East ceased to exist as a defined area and has since been considered part of the Middle East. Today the Middle East consists of Afghanistan, Bahrain, Cyprus, Iran, Iraq, Israel, Jordan, Kuwait, Lebanon, Oman, Qatar, Saudi Arabia, Syria, Turkey, the United Arab Emirates, and Yemen, although as a political concept, the region also includes Egypt in Africa.

The region is known for its so-called Fertile Crescent, where moderate rainfall augmented by extensive irrigation systems around the Tigris and Euphrates rivers gave rise to early Middle Eastern civilization. Much of the Middle East's climate in modern times, however, is harsh: temperatures sometimes reach 130°F in the summer or drop to −35°F (in central Turkey and Afghanistan) in the winter. Until recently the area still reflected an Old Testament−era economy, with peasant farmers and nomadic herders, who still live simply amid the great wealth of the Middle East nations that control profits extracted from oil making up about two-thirds of the world's reserves.

AFGHANISTAN

With numerous rivers and streams, Afghanistan possesses many fertile plains and valleys. Even so, southern Afghanistan is primarily desert, while the central and northern regions are dominated by 15,000- to 25,000- foot ranges of the Hindu Kush Mountains. Rivers include the Helmand in central and southwest Afghanistan; the Hari Rud, Harut Rud, Farah Rud, and Khash Rud in the east; and in the north the Amu Darya, which forms part of the boundaries with Tajikistan and Turkmenistan, members of the Commonwealth of Independent States (formerly the Union of Soviet Socialist Republics).

Afghanistan once formed part of the Persian empire; it became part of the Mogul empire of India near the end of the fifteenth century and was seized by Persia in the first half of the seventeenth century and consolidated as a separate entity. By the late 1700s, the Afghan empire included eastern Persia, modern Pakistan's Baluchistan and Punjab, and Kashmir, now divided between Pakistan and India. The boundaries with India were settled in 1893, and with Russia in 1895.

BAHRAIN

Known as Tylos in ancient times, Bahrain consists of an archipelago in the Persian Gulf.

CYPRUS

Once famous for its rich deposits of copper, Cyprus has been a valued prize as far back as the days of the Phoenicians and ancient Greeks, who colonized the island. Its Greek name, Kypros, comes from the Latin word *cuprum*, which means "copper."

IRAN

Formerly known as Persia, the Islamic Republic of Iran consists of 33,600 square miles of plateaus and mountains, especially

127

the Elburz Mountains in the north along the Caspian Sea. The Hindu Kush Mountain ranges reach into northeastern Iran, and many ranges in the west rise higher than 10,000 feet. Eastern Iran is dominated by the Plateau of Iran.

The ancient civilizations of Elam, Media, and Persia inhabited the plateau before 2000 BC. In the sixth century BC, Cyrus the Great founded the Persian empire, extending from the Indus River in India to the Mediterranean Sea and from the Caucasus (between the Black and Caspian seas) to the Indian Ocean. Following its conquest by Alexander the Great, Persia was ruled by outsiders before the Neo-Persian empire was founded in the third century AD, but it was conquered by Asia's Mongols in the thirteenth century. Modern Persia was founded at the beginning of the sixteenth century but lost the Caucasus to Russia in the nineteenth century. Persia was officially renamed Iran in 1935.

IRAQ

Mostly level country drained by the Tigris and Euphrates rivers, which merge to form the Shatt-al-Arab River for another 120 miles, Iraq boasts fertile river regions, including most of Mesopotamia. The Syrian Desert covers the western region, while the northeastern Kurdistan region is mountainous.

In ancient times, in the region of lower Mesopotamia, Iraq Arabi comprised the cities

Basra, Baghdad, and adjoining provinces in Iraq and was nearly coextensive with Babylonia. Modern Iraq was established after World War I out of former Turkish territory and remained under British mandate until achieving independence in 1932. In 1990, Iraq illegally annexed Kuwait, to which it had long laid territorial claim, but lost it again following the Gulf War in early 1991.

ISRAEL

Established in 1948, modern Israel consists of nearly 8,000 square miles, dominated by hills (the highest just 3,700 feet) in the north and the Negev Desert in the south. The coastal plain, with a maximum width of 20 miles, and the Plain of Esdraelon are less than 300 feet below sea level. Bounded by the Mediterranean Sea on the west, Israel has a port in the extreme south on the Gulf of 'Aqaba.

As recommended by a special committee of the United Nations following World War II, Palestine was partitioned between Jews and Arabs, leading to the establishment of Israel in May 1948, intensifying the state of civil war in Palestine and precipitating continued animosities between Israel and neighboring Arab states. In July 1948, Egypt annexed the Negev region, which Israel regained in offensive actions the following December. Also in 1948, Israel occupied Jerusalem and the following year declared it a part of the state of Israel. An armistice agreement allowed Israel to retain Jerusalem, but it yielded to Egypt the coastal Gaza Strip, which it had occupied in its Sinai campaign during the 1956-1957 war with Arab countries. In 1967, Israel occupied and annexed Golan Heights, which adjoins Syria, and parts of Jordan west of the Jordan River, known as the West Bank.

JORDAN

Like Iraq, Jordan was created out of former Turkish territory in 1921; it was declared an independent state two years later but remained under British mandate until 1948, when the mandate was revoked and Jordan was declared an independent kingdom, officially named Transjordan. On signing the armistice agreement following its 1948–1949 war with Israel, Transjordan renamed itself Jordan, or officially Hashemite Kingdom of Jordan. In 1950, Jordan formally annexed the West Bank, land west of the Jordan River, which Israel occupied and annexed after the Arab-Israeli War in 1967.

KUWAIT

A desert territory comprising 6,200 square miles between Iraq and Saudi Arabia, Kuwait was occupied and annexed by Iraq in 1990, regaining its independence following the Gulf War in early 1991.

LEBANON

Largely mountainous, Lebanon comprises about 3,950 square miles, with the Lebanon Mountains cutting through the central region.

OMAN

Formerly named Musqat and Oman, Oman is a sultanate led by rulers who take the title "emir." Northern Oman is mountainous, while western Oman is desert. The coastline on the Arabian Sea runs for about 1,000 miles.

QATAR

The Sheikdom of Qatar comprises 4,400 square miles of low hills and desert.

SAUDI ARABIA

A plateau region, the Kingdom of Saudi Arabia has an average elevation of 2,500 feet, with 7,000- to 10,000-foot highlands in the west. Its 874,000 square miles include the Nefud Desert in the north and the Rub'al Khali Desert in the south.

SYRIA

The northern part of the Syrian Desert dominates southern Syria, while the upper part of the Plain of Mesopotamia lies in the northeast. Central Syria is a plateau, its elevation ranging from 1,500 feet to 4,500 feet. The Jebel er Ruwaq Mountains and the northern extension of the Lebanon mountain range lie to the west.

The Euphrates River flows through eastern and central Syria. The country covers 71,500 square miles.

TURKEY

A mountainous country with extensive plateau covering central Asia Minor (the peninsula formerly the northwestern extremity of the Middle East), Turkey comprises more than 300,000 square miles, with the Black Sea to the north and the Mediterranean Sea to the south. Its highest peak, Mt. Ararat, is where the biblical ark is believed to have come to rest after the Great Flood. Its rivers include the upper courses of the Tigris and Euphrates. About three percent of the country lies in Europe, on the Balkan Peninsula.

Following the defeat of the Ottoman Empire, the former sultanate that spread across the Middle East and into Europe and Africa, the nationalist government finally proclaimed the Republic of Turkey in 1923. The capital was moved to Ankara from the Ottoman capital Istanbul (formerly Constantinople, the name given in AD 330 by Constantine, the emperor of the Byzantine empire).

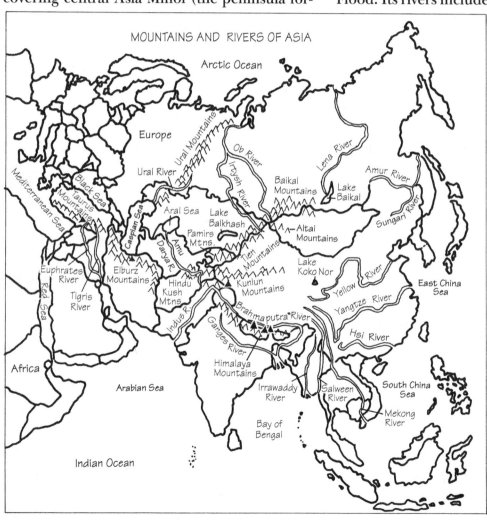

MOUNTAINS AND RIVERS OF ASIA

UNITED ARAB EMIRATES

A federation of seven states, the United Arab Emirates extends from Qatar to the Gulf of Oman, with a 400-mile coastline. While the federation comprises 30,000 square miles, only one of the states is larger than 1,500 square miles.

YEMEN

Two republics—the Yemen Arab Republic and the People's Democratic Republic of Yemen, were unified in 1990. San'a, the former capital of the Yemen Arab Republic, became the capital of the unified nation.

Monsoons, Cyclones, and Tornadoes

Asia's climate is characterized by distinct wet and dry seasons called "monsoons." (The word *monsoon* comes from the Arabic word *mawsim*, meaning "season.") During the summer monsoon, winds and heavy rain predominate. When trade winds from the northeast and southeast collide around the equator, the air rises, the atmospheric pressure falls, and condensation forms clouds that produce heavy precipitation.

While monsoons can be found in all tropical areas under other names, they are typically associated with southern Asia. During the summer months, the heat over Asia creates low pressure over most of the continent, while warm, moist air is drawn from the Indian Ocean, resulting in rain-producing clouds. In autumn there is less contrast between land and sea temperatures, and the areas of low and high pressure reverse. Cooler, drier air blows down from the Himalayas and persists until spring.

Agriculture in much of Asia depends on the monsoon season, and while the rainy season generally occurs from April to October, it varies. When the monsoons arrive late, less rain falls, often resulting in a tremendous loss of crops and bringing famine. In India monsoons provide the desert-like subcontinent with sufficient moisture to support a huge population.

Cyclones can also result when atmospheric pressure falls. They, too, can occur elsewhere—they're called "hurricanes" in the Caribbean Sea and the Gulf of Mexico and typhoons in the western Pacific Ocean. Asiatic cyclones develop in the Bay of Bengal, a division of the Indian Ocean surrounded on three sides principally by the mainland countries of India, Bangladesh, Myanmar (Burma), and Thailand. A cyclone is initiated when a wave develops on a warm front (the boundary between cold and warm air masses) with a cold front.

In April 1991 a cyclone traveled northwest across the Bay of Bengal; its strong winds and mighty walls of water struck the southern coast of Bangladesh and killed more than 100,000 people.

Following on the heels of that huge killer cyclone, a tornado further damaged the devastated

Asia Facts

· With nearly 287,000 square miles, Indonesia's Borneo is the world's third largest island.

· Dating back thousands of years BC, the ancient cultures of China, India, and the Indus Valley in southern Asia; and Arabia, Assyria, Babylon, the Levant, Mesopotamia, and Persia in the Middle East were the nurseries of civilization and of the world's major religions, including Hinduism, Buddhism, Islam, Christianity, and Judaism.

· Mostly in the southeast of the Mongolian People's Republic and northern China (Inner Mongolia), the Gobi Desert comprises about half a million square miles. While the western Gobi is entirely covered by sand, the rest is part of the great Eurasian steppe, the treeless grassland that extends from Ukraine, through southeastern Europe and central Asia, to Manchuria. The steppe is equivalent to the North American plains and the South American pampas.

· More than 5,700 feet deep, Siberia's crescent-shaped Lake Baikal is the world's deepest lake. Fed by more than 300 rivers and streams, it is also Asia's largest freshwater lake, encompassing about 12,160 square miles.

country. Meteorologists do not fully understand how the destructive columns of rapidly rotating air are formed, but tornadoes (or "twisters") are also associated with low atmospheric pressure.

Major Mountain Ranges

Located in central Asia, north of the Indus and Brahmaputra rivers, the Himalayas (pronounced hi-*mahl*-yaz) dominate the continent of Asia. More than 30 peaks stand higher than 28,200 feet—and are still rising as a result of the Indian-Australian plate's continued push against the continent of Asia.

With Alpine pastures, tropical forests, coniferous woods, bamboo, and rhododendrons in their lower regions, the snow- and ice-covered mountains rise toward the north, extending for more than 1,500 miles from the Pamirs in the northwest to the borders of China and the Indian state of Assam in the east. Three main ranges—the Outer, Middle, and Inner Himalayas—become five ranges in the Indian state of Kashmir: the Lesser and the Great Himalayas, the Zaskar Range, the Ladakh Range, and the Karakorams.

· Mt. Everest, the highest mountain in the world, is located on the

Nepal-Tibet border and is over 29,028 feet tall.

- K2 (also known as Godwin Austen) is in Kashmir, India: 28,250 feet.
- Kanchenjunga on the India-Nepal border: 28,208 feet.
- Makalu on the Nepal-Tibet border: 27,824 feet.
- Dhaulagiri in Nepal: 26,810 feet.
- Cho Oyu in Tibet: 26,750 feet.
- Nanga Parbat in Kashmir, India: 26,660 feet.
- Annapurna in Nepal: 26,504 feet.
- Gasherbrum in Kashmir: 26,470 feet.
- Broad in Kashmir: 26,400 feet.

Major Volcanoes

Asia is a hotbed of volcanic activity. Japan has numerous active volcanoes. In June 1991 another volcano in the area, Mt. Unzen (about 4,500 feet), erupted for the first time in about 200 years.

On Honshu, the island where Japan's capital is located, snow-covered Mt. Fuji, also called Fujiyama and Fujisan, is considered a sacred mountain and hasn't erupted since 1707.

On the Philippines in June 1991, Mt. Pinatubo, on the island of Luzon, erupted for the first time in 600 years; volcanic ash covered the United States' Clark Air Base ten miles to the east, making the base unusable and forcing the evacuation of military personnel and their fami-lies. The eruption also threatened to cause the evacuation of the United States' Subic Bay Naval Base to the south.

Indonesia's island of Sumatra has 12 active volcanoes. The 1883 eruption of Krakatoa, or Krakatau, a small volcanic island in the Sunda Strait, was the most violent in recent world history, causing a 50-foot tidal wave and killing 36,000 people in western Java. Dust, ash, and smoke rose 17 miles, and the explosion was heard as far west as Turkey.

Essential Statistics

Westernmost point: Cape Baba, Turkey, 26° 04' east longitude.

Easternmost point: Dezhneva Cape, the Russian Federation, on the Bering Strait, 169° 40' west longitude.

Northernmost point: Cape Chelyuskin, the Russian Federation, 77° 45' north latitude.

Southernmost point (mainland): Cape Piai, Malaysia, 1° 15' north latitude.

Highest point: Mt. Everest (about 29,000 feet), on the Himalayan border between Nepal and Tibet, also the highest peak on Earth.

Lowest point: The Dead Sea (about 1,290 feet below sea level), in both eastern Israel and western Jordan.

Quiz

(answers, p. 136)

1. What is the largest country completely on the continent of Asia? What is the second largest?

2. In December 1984, Bhopal was the scene of a major industrial disaster when poisonous gas escaped from the Union Carbide factory, killing some 2,500 people and leaving 100,000 homeless. In what country is Bhopal?

3. Identify the following ancient regions by their modern names:
   a. Cathay
   b. Persia
   c. Mesopotamia
   d. Assyria

4. In the Rodgers and Hammerstein musical *The King and I*, the king rules the Kingdom of Siam. By what name is Siam known today?

5. Madras plaids (made with dyes that run when the fabric is washed) are named for an area in southern Asia. Similarly, dungarees derive their name from Dugri, a region in the same country as Madras. In what country are Madras and Dugri located?

6. Tokyo became the capital of Japan in 1868. In English the new capital's name uses the same letters as those that make up the name of the former capital. What was Tokyo's former name?

7. What country is nicknamed "The Roof of the World"?

8. In what country is Bethlehem, considered by Christians to be the birthplace of Jesus Christ?

9. To what countries would you go to see the following:
   a. Mt. Everest
   b. the Taj Mahal
   c. Tiananmen Square
   d. Mt. Fuji
   e. Mt. Ararat
   f. the Great Wall

10. You are going to tour southern and southeastern Asia. Arriving in Hong Kong, you'll continue on to Borneo, Myanmar, Sri Lanka, Thailand, and Vietnam. Plan your itinerary so that you proceed in an orderly fashion and can depart from Dum Dum Airport on your flight home.

11. Closed to foreigners in the nineteenth century, what Himalayan capital is known as "The Forbidden City"?

12. The area west of the Jordan River, occupied by Israel since the 1967 Six-Day Arab-Israeli War, was once part of Palestine, which no longer exists. By what name is the area commonly identified today?

13. In 1975, Saigon, the capital of South Vietnam, was renamed. What is it now called?

14. Where is Mecca, the holy city of Islam?

15. What modern country is the site of ancient Troy, the city where, according to Christopher Marlowe's *Doctor Faustus* (1604),

Helen's face "launched a thousand ships"?

16. In what modern country were the Hanging Gardens of Babylon, one of the Seven Wonders of the Ancient World, once located?

17. In what city would you find the Wailing Wall?

18. Where were Sodom and Gomorrah, the wicked biblical cities destroyed by fire and brimstone as divine judgment?

19. Determine the latitude and longitude of the following cities:
    a. Bangkok
    b. Calcutta
    c. Jerusalem
    d. Manila
    e. Mecca
    f. Shanghai
    g. Tokyo

ANSWERS

1. China is the largest, India the second largest. (The Commonwealth of Independent States is not completely on the continent of Asia.)
2. India
3. a. China
   b. Iran
   c. Iraq
   d. Iraq
4. Thailand
5. India. Madras is the main port on the southeast coast. Dugri is a suburb of Bombay.
6. Kyoto was founded in the eighth century and served as Japan's capital nearly 1,000 years. Tokyo was founded in the twelfth century as the village of Edo.
7. Tibet received the nickname because it is home to the world's highest mountains, the Himalayas.
8. Officially, Bethlehem lies on Jordan's side of the Israeli-Jordanian border, established by the 1948 partition of Palestine into the State of Israel and the western portion of the Kingdom of Jordan. Bethlehem, along with the area known as the West Bank, has been under Israeli administration since the 1967 Arab-Israeli War.
9. a. Nepal; Tibet
   b. India
   c. This site of the June 1989 student demonstrations and government crackdown is located in China's capital, Beijing.
   d. Japan
   e. The supposed landing place of biblical Noah's Ark is located in eastern Turkey, close to the border with Iran and the Commonwealth of Independent States.
   f. During the third century BC, 300,000 men, mostly criminals, constructed the 2,000-mile defensive barrier to protect China against invasion from the north. Today China's northern border is actually north of the Great Wall.
10. From Hong Kong, fly to Borneo, north to Vietnam, then to Thailand. Cross the border into Myanmar, and from there fly to Sri Lanka, off the southeast coast of India, then to India, where you'll catch your plane home.
11. Lhasa, Tibet
12. The West Bank
13. Ho Chi Minh City
14. Saudi Arabia
15. Turkey
16. Iraq
17. Jerusalem
18. In ancient Palestine. Today their original sites may be submerged by the Dead Sea.
19. a. Bangkok, Thailand: 13° 45' north latitude, 100° 30' east longitude

b. Calcutta, India: 22° 34' north latitude, 88° 24' east longitude

c. Jerusalem, Israel: 31° 47' north latitude, 35° 15' east longitude

d. Manila, Philippines: 14° 35' north latitude, 20° 57' east longitude

e. Mecca, Saudi Arabia: 21° 29' north latitude, 39° 45' east longitude

f. Shanghai, China: 31° 10' north latitude, 121° 28' east longitude

g. Tokyo, Japan: 35° 40' north latitude, 139° 40' east longitude

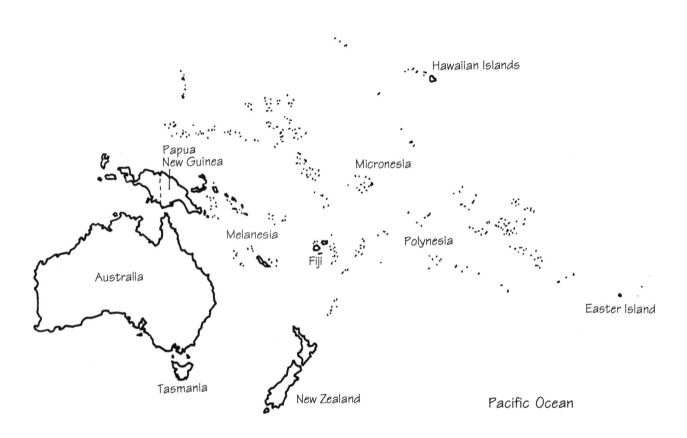

Pacific Ocean

Hawaiian Islands

Papua
New Guinea

Micronesia

Melanesia

Fiji

Polynesia

Easter Island

Australia

New Zealand

Tasmania

Pacific Ocean

OCEANIA

There are some 25,000 islands in the Pacific Ocean, most of them lying in the Southern Hemisphere, most of them small, many uninhabited, and most of them coral atolls or extinct volcanoes. Geographers collectively refer to the South Pacific's populated islands as Oceania. Although the South Pacific technically falls south of the equator, which divides the globe into the Northern and Southern hemispheres, Oceania also includes the Hawaiian Islands, which lie across the Tropic of Cancer in the Northern Hemisphere.

Oceania is divided into three subgroups: Polynesia, Melanesia, and Micronesia. But like Tahitians, Hawaiians and New Zealand's Maoris are Polynesians. Many anthropologists racially categorize Australia's very dark-skinned aborigines as Melanesian (from *melas,* meaning "black"), but because their numbers are small compared to the European Australians, the continent of Australia is not included geographically in Melanesia. Today English is spoken on most of the islands, but among the island groups the primary language tends to be the native tongue of the individual group. It's really not as confusing as it appears. A closer look should help.

AUSTRALIA

Australia is the only country to occupy an entire continent, which also bears the same name. Nearly 3 million square miles, including the island of Tasmania, the country is almost as large as the contiguous United States, with the city of Perth, on the west coast, just about as far from Sydney, in the southeast, as Los Angeles is from New York.

Australia is bounded by the Indian Ocean on the west and south, and by the Coral and Tasman seas of the Pacific Ocean on the east. The two oceans meet north of Australia in the Timor and Arafura seas. On the eastern side of the continent, the Great Dividing Range extends from jungle-covered Cape York in north-

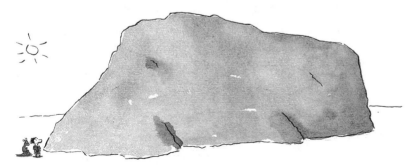

Australia contains many plants (like the eucalyptus) and animals that are indigenous nowhere else in the world. The continent is especially known for its marsupials—kangaroos, koalas, raccoonlike Tasmanian devils, and bearlike wombats. Other unique species include the platypus, the dingo (wild dog), and barking lizards.

ern Queensland to the southern coast of Victoria; the mountain range includes the Australian Alps along the southeast coast. West of the Great Dividing Range, one of the world's greatest sheep populations grazes on grassy steppes. Australia provides a third of all the world's wool.

Beyond this, the vast arid plateau known as the Australian Outback stretches across most of the continent. It is marked by the Great Victoria Desert and the Great Sandy Desert, as well as the Gibson, Simpson, Tanami and other smaller deserts, and rises to rolling hills on the west coast. The highest point is Mt. Kosciusko (about 7,300 feet) in the Great Dividing Range. Ayers Rock, literally in the middle of nowhere, towers 1,100 feet above the otherwise flat Outback.

The rainfall is heaviest in the northeast. In the southeast the Murray River rises in East Victoria, flows for about 1,600 miles, and empties into the Indian Ocean. The largest of Australia's salt lakes is Lake Eyre, a lagoon of 3,600 square miles with a maximum depth of four feet. Like the island's aborigines at the time of Australia's discovery by Europeans, today's population is concentrated in the coastal regions, principally in the east.

The island state of Tasmania, with its central highland, lies 150 miles south of the state of Victoria in the southeast, separated from the mainland by the Bass Strait. Australia's island territories include Norfolk, Cocos, Christmas (225 miles south of Java), Heard, and McDonald islands, Ashmore and Cartier Island, and the Coral Sea Islands Territory northeast of the Great Barrier Reef. Designated a protected marine park in 1983, the 1,250-mile-long Great Barrier Reef off the northeast coast is the world's largest deposit of coral.

Although they have never lived on more than about one-tenth of Australia's entire 3 million square miles, the Australian aborigines probably arrived on the island continent from Southeast Asia across a land bridge that once connected Australia and New Guinea. Even today Australia's generally European population is small in comparison to the country's area, with the people concentrated in a handful of modern cities.

A Portuguese navigator, Manuel Godhino de Eredia, probably sighted Australia in 1601. The Spaniard Luis Vaez de Torres did so soon after; then came the Dutch in 1606. By the middle of the century, they had explored the northern and western regions, claimed it, and named it New Holland. More than a century later, British navy captain James Cook discovered Botany Bay, south of modern-day Sydney on the east coast, and named the land New South Wales. No one realized that New Holland and New South Wales were actually on the same landmass until the continent was circumnavigated in the early nineteenth century. In 1829 the British claimed the entire continent, which they had referred to as "terra australis incognita," and soon after began calling it "Australia."

The Dutch never rushed to colonize New Holland, a harsh land they considered uninhabitable. The British, however, found the remote New South Wales, isolated from other colonies and territories, appealing as a penal colony. In the late 1780s they established Port Jackson, Australia's first settlement for convicts, near what is now Sydney.

Deporting convicts overseas was not a new idea, and Britain had even sent them to its American colonies. The practice was certainly not confined to Great Britain. The French, for example, opened penal colonies on islands off the northern coast of South America and on the island of New Caledonia (see Melanesia). At the time of Australia's settlement, Britain's criminal shipments to North America had been curtailed by the American Revolution, and jails in England were becoming overcrowded. Australia, however, turned out to be the biggest penal colony the world has ever known.

The last deported convict reached Australia in the mid-nineteenth century. By that time, free immigrants, who began settling in Australia in 1816, had established the colonies of Tasmania (then called Van Diemen's Land), Western Australia, South Australia, Victoria, and Queensland and the Northern Territories. These colonies, along with New South Wales, became states and, in 1901, were federated into the Commonwealth of Australia.

Following the enactment of the Australian Colonies Government Act of 1850, which granted self-government to Australia, the country adopted its "whites only" immigration policy, beginning with the exclusion of Chinese immigrants. The racially discriminatory restrictions were not abandoned until 1973, following international pressure.

In 1900 the British Parliament passed an act permitting the federation of the Australian states to form into a commonwealth, which wasn't finalized until 1901. The site of the federal capital, Canberra, was chosen in 1908. Today the Australian Capital Territory of 925 square miles, encompassing Canberra, is separated from the rest of New South Wales, much as the District of Columbia encompasses the United States capital of Washington and is separated from the state of Maryland.

NEW ZEALAND

New Zealand's 103,700 square miles includes North Island (more than 44,000 square miles), South Island (about 58,300 square miles), and Stewart Island (about 675 square miles), which is 1,250 miles southeast of Australia. A number of the scattered smaller island groups—including the Chatham Islands, the Auckland Islands, Campbell Island, the Antipodes Islands, and the Kermadec Islands—range from the tropical to the antarctic in climate. Foveaux Strait separates

Stewart Island from the southern coast of South Island.

Separated by Cook Strait, the two main islands are hilly and mountainous, with fertile plains on the east coasts, notably the Canterbury Plains on South Island. Containing numerous glaciers, the Southern Alps extend over almost the entire western coast of South Island, with 15 peaks more than 10,000 feet. The highest point is Mt. Cook (about 12,350 feet). Lakes include the Wakatipu, Wanaka, and Te Anau; rivers include the Wairau, Rangitata, Waitaki, and Clutha.

A volcanic plateau containing hot springs and geysers lies in the center of North Island, as does Lake Taupo. Rivers include the Waikato, Wanganui, Rangitaiki, and Rangitikei. The capital, Wellington, is the world's southernmost capital city. Other major cities are Auckland on North Island and Christchurch on South Island.

The Maoris, a Polynesian group from the eastern Pacific, arrived in New Zealand before the fourteenth century. Dutch navigator Abel Tasman discovered New Zealand in 1642, naming it after a province in the Netherlands. He would have landed had the Maoris permitted, but they didn't. Captain James Cook circumnavigated and visited the islands in 1769 and was followed by the first settlers—missionaries and whalers who used the islands as their base during whaling season.

Formal settlements weren't established until 1840, following the Treaty of Waitangi, by which tribal leaders ceded lands to the British, who then declared New Zealand a British colony. But relations between the two peoples weren't always peaceful, especially when the British began expanding their territory. The first British-Maori war lasted from 1843 to 1848; the second war lasted from 1865 to 1870. The capital was moved from Auckland to Wellington in 1865.

New Zealand is now an autonomous member of the British Commonwealth of Nations. Four of the 95 members of its parliament are elected directly by the native Maoris.

Polynesia

Polynesia is a region of many islands (thus the *poly*, meaning "many") with a common history and culture. It includes Tonga, Easter, Western Samoa, American Samoa, Pitcairn, Tuvalu, and French Polynesia island groups concentrated east of Indonesia. New Zealand's Cook Islands, Tokelau Islands, and Niue also make up Polynesia, as do New Zealand itself (which lies south of the Tropic of Capricorn) and the Hawaiian Islands, north of the Tropic of Cancer. Polynesia ranges from volcanic islands (actually the tops of volcanoes) covered with tropical vegetation and receiving more than 100 inches of rain every year, to coral atolls (ring-shaped coral islands), many of them barely above sea level, with only a few palm trees and experiencing persistent drought. The greater number of the inhabitants are Polynesians, perhaps related to the Malays of Southeast Asia, but many are of mixed origins.

The center of Polynesian settlement was probably Samoa (now American Samoa and Western Samoa). Long before Europeans began arriving at their islands, the Polynesians navigated great expanses of the Pacific in canoes as long as 150 feet. They used maps, colonized uninhabited islands they encountered, and engaged in inter-island trade. Although Polynesia conjures up romantic images of a Pacific paradise, many Polynesians suffered starvation, especially on the smaller islands, and many died at sea.

Despite the expansiveness of the Polynesian islands, however, the traditional culture—its language, architecture, and art—is consistent from

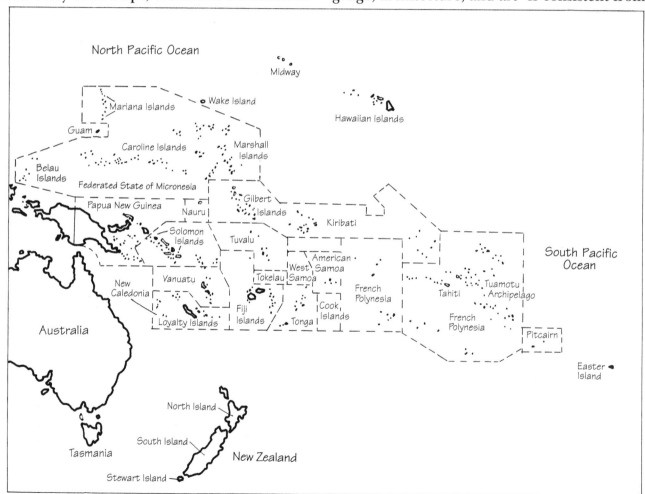

one island to the next. Even so, modern Polynesian societies, especially in the Polynesian-American-Asian Hawaiian Islands, have adopted European as well as Asian culture.

HAWAII

Hawaii is actually a 1,500-mile-long chain of islands, the tops of submerged volcanic mountains, with a total land area of more than 6,400 square miles. The U.S. state of Hawaii officially includes the eight major islands—Hawaii, Oahu, Maui, Kauai, Niihau, Lanai, Molokai, and Kahoolawe. The more than 100 northwestern Hawaiian Islands (except for Midway) are an administrative part of Hawaii. The island of Kauai is the wettest spot in the United States, with an annual rainfall of 444 inches. The highest peak is Mauna Kea (nearly 13,800 feet) on Hawaii. The state capital, Honolulu, is on the island of Oahu.

Hawaii was first settled around AD 500 by Polynesians arriving from the south. The first known European to visit the islands was the British captain James Cook, who named the chain the Sandwich Islands in the late eighteenth century. Frequented by American whalers and missionaries during the nineteenth century, the Hawaiian Islands remained a native kingdom throughout most of that century, with its independence recognized by Great Britain, the United States, and France. In the 1890s, following the last two kings' misrule of the islands and the deposition of the queen, a republic was set up, headed by Sanford Dole, of the pineapple-growing family. Strongly backed by the republic's government, the United States annexed the islands and established them as a territory in 1900. The eight main islands became a U.S. state in 1959.

Island Facts

Visited by the British captains James Cook and William Bligh, Tahiti has appealed to Herman Melville, Charles Darwin, and Paul Gauguin, whose Tahitian-inspired art is among his best known. In more recent times, Tahiti's best-known resident has been actor Marlon Brando.

Pitcairn Island was discovered around 1767, but wasn't settled until more than 20 years later and then by mutineers led by Fletcher Christian from HMS *Bounty*. In subsequent years the population was moved to Tahiti, in present-day French Polynesia, and later to Norfolk Island, east of Australia. Some of the population, however, eventually returned, and their descendants (many from the *Bounty* crew) constitute the present population.

Germany claimed the Marshall Islands around 1885. Japan invaded them during World War I and held them until World War II, when the United States invaded them. The islands became part of the U.S. Trust Territory of the Pacific Islands in 1947, established by the United Nations in 1947. Between 1946 and 1958, atom and hydrogen bomb testing was conducted on two of the Marshall Islands, Bikini and Eniwetok.

Melanesia

The islands of Melanesia are the most populated of the three island groups in the South Pacific. The world's third largest island, New Guinea, is in this region.

FIJI

The Republic of Fiji comprises more than 800 islands, about 100 of them inhabited. The larger islands, mostly mountainous and volcanic, include the principal islands, Vanua Levu and Viti Levu. By law, native Fijians own 83 percent of the land in communal villages.

NEW CALEDONIA

New Caledonia, a French territory northeast of Sydney, Australia, includes the Loyalty Islands, Huon Islands, Chesterfield Islands, Belep Islands, and the Isle of Pines. The largest island, New Caledonia, has one of the world's largest nickel deposits.

PAPUA NEW GUINEA

Papua New Guinea constitutes the eastern half of the island of New Guinea, which it shares with Irian Jaya, a province of Indonesia. Numerous isolated tribes, speaking 750 mutually unintelligible languages, inhabit the mountainous, heavily forested interior, which has only recently been explored.

SOLOMON ISLANDS

The Solomon Islands, independent since 1978, comprise a group of largely volcanic islands east of Papua New Guinea and include the islands of Guadalcanal, Choiseul, Malaita, New Georgia, San Cristobal, Santa Isabel, and numerous smaller islands.

Micronesia

The islands of Micronesia were organized into the U.S. Trust Territory of the Pacific Islands in 1947 under the guidance of the United Nations and are currently administered by the U.S. Department of the Interior. The area includes some 2,000 islands and atolls extending over some 3 million square miles. The land area, however, comprises little more than 1,000 square miles and includes three major archipelagos: the Carolines, the Marianas, and the Marshalls. The Carolines include the Federated State of Micronesia and the Republic of Belau. Other Micronesian islands include Guam and the Kiribati (Gilbert) Islands.

GUAM

Guam, the largest and southernmost of the Mariana Islands, is a self-governing territory of the United States and not part of the United States' Marianas trusteeship. Self-government and American citizenship were granted in 1950. The capital is Agana.

Quiz

1. The Japanese attack on Pearl Harbor occurring on December 7, 1941, led to the United States' entry into World War II. Where is Pearl Harbor?

2. To what general island group do New Zealand and Hawaii belong?

3. What is the only country that covers an entire continent?

4. The island of Guadalcanal was the site of numerous naval and land battles between Allied forces and the Japanese during World War II. To what Pacific islands does Guadalcanal belong?

5. What island is not part of the state of Hawaii, although it is one of the Hawaiian Islands and belongs to the United States?

6. What country's citizens are called Kiwis?

7. On which of the Hawaiian Islands can you find Honolulu, the capital of Hawaii?

8. What city boasts the only royal palace in the United States?

9. By what name are the Sandwich Islands known today?

10. What is the world's largest coral reef?

11. Which one of the Hawaiian Islands is usually referred to as the Big Island?

12. What country was first known as New Holland?

13. What island was settled by British sailors who mutinied against William Bligh, master of HMS *Bounty*?

14. What Hawaiian mountain boasts the world's largest active volcano?

15. On what Pacific island can you find Volcano National Park?

16. What country's first European settlers were primarily convicts?

ANSWERS

1. Oahu, in the U.S. state of Hawaii
2. Polynesia
3. Australia
4. The Solomon Islands
5. Midway Island
6. New Zealand
7. Oahu
8. Honolulu
9. The Hawaiian Islands
10. The Great Barrier Reef, off the coast of Australia
11. The island of Hawaii
12. The Dutch were the first to land on Australia, which they named New Holland. The British called it New South Wales. The name was officially changed to Australia in the first quarter of the nineteenth century.
13. Pitcairn Island
14. Mauna Loa
15. The island of Hawaii
16. Australia

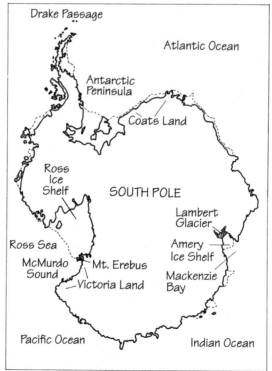

POLAR REGIONS

The earth takes approximately 365 days (one year) to complete one full orbit around the sun, but only 24 hours to rotate about its own axis, an imaginary straight line that cuts through the planet's surface. The planet's geographical poles, the North Pole and the South Pole, lie at the diametrically opposite ends of this axis. The North Pole lies beneath the Arctic Ocean's permanent ice pack; the South Pole, on the continent of Antarctica.

Encircling the North Pole, the Arctic Circle is an imaginary line around the earth at 66° 17' north latitude—although the Arctic region is often defined as the area north of 70° north latitude. Within the Arctic Circle, the sun doesn't set during the northern summer solstice (June 21). During the northern winter solstice (December 21), the sun doesn't rise.

In the Southern Hemisphere, at 66° 30' south latitude, the Antarctic Circle surrounds the South Pole. Within this region, the sun doesn't set during the southern summer solstice (December 21) and doesn't rise during the southern winter solstice (June 21), the exact opposite of the Northern Hemisphere's Arctic region.

The Antarctic

Lying almost completely within the Antarctic Circle, Antarctica comprises 5.5 million square miles and is surrounded by the ice-filled Antarctic Ocean, which is actually the southern extensions of the Atlantic, Indian, and Pacific oceans. The Drake Passage separates the continent from South America by 690 miles at its narrowest point. The Transantarctic Mountains, a mountain range running from Victoria Land to Coats Land, divides the continent into Greater (or East) Antarctica and Lesser (or West) Antarctica, which includes the Antarctic Peninsula. Central Antarctica is dominated by the South Polar Plateau, which rises from the continent at the southernmost part of the world and surrounds the South Pole. The continent has no lakes or rivers.

The depth of the ice cap that covers Antarctica averages about a mile and reaches a maximum depth of about

149

14,000 feet in Lesser Antarctica and about 9,000 feet in Greater Antarctica. Glaciers slowly move toward the coasts, their quantity and power often so great that they push beyond the landmass into the sea, creating an ice shelf over large areas and forming high ice cliffs around the redefined coastline. The Lambert Glacier, for example, feeds the Amery Ice Shelf on Mackenzie Bay, and the Beardmore Glacier feeds the Ross Ice Shelf on the Ross Sea.

Some places on the continent, however, are free of ice. Peaks protrude through the ice cap, mostly in Lesser Antarctica. And part of the Antarctic Peninsula (formerly called Palmer Peninsula) is a bare, black-rocked volcanic plateau. The Antarctic Peninsula, in fact, has yielded important fossil and mineral finds.

Although the presence of coal indicates that Antarctica once had a warm climate, the conti-

nent now has the world's harshest climate. Temperatures rarely exceed the freezing point on sunny days. Winter winds average about 48 miles per hour, but 150-mile-per-hour gales can blow up unexpectedly and continue for 24 hours at a stretch. And in a whiteout, when driving snows obliterate everything from sight, it's possible to get lost only a few feet from a camp hut.

The writings of Ptolemy and other ancient Greeks referred to a great land to the south that many researchers believe was Antarctica. Eighteenth-century European navigators approached the continent as they explored the Southern Hemisphere and circumnavigated the globe, but no one laid claim to discovering Antarctica until the nineteenth century. In 1820, Nathaniel Palmer of the United States discovered Palmer Peninsula (now called Antarctic Peninsula) without realizing that it was a continent. Although British, Norwegian, German, Belgian, and other expeditions continued the discovery and scientific study of the mainland, no one really considered Antarctica to be a single landmass and one continent until the Ronne Antarctic Research Expedition (1946–1948), headed by Finn Ronne, revealed that no strait connected the Weddell and Ross seas.

On December 14, 1911, Norwegian explorer Roald Amundsen was the first man to reach the South Pole. On Amundsen's heels, British explorer Robert Scott reached the pole on January 18, 1912, but the British team didn't survive.

Science in the Antarctic

Antarctica has been heralded as a laboratory for the space program ever since the signing of the Antarctic Treaty, which foreshadowed the 1967 outer space treaty. The isolated, intimate living conditions in Antarctica, for example, resemble those that astronauts confront during long journeys in outer space. And planners of the Mariner missions to Mars, in fact, used Antarctic valleys as test models.

Reserving the continent for science, the Antarctic Treaty provides for international cooperation in scientific research and prohibits military bases, nuclear and other weapons tests, and disposal of nuclear waste. Today 39 nations are treaty members, but only 26 of them have full voting rights on matters related to Antarctica. Of the original 13 members—Argentina, Australia, Chile, France, Germany, Japan, New Zealand, Norway, Poland, South Africa, Russia, the United Kingdom, and the United States—seven still claim territorial rights to the region: Argentina (74° west longitude to 25° west); Australia (Enderby Land, Wilkes Land, George V Coast, part of Oates Coast); Chile (90° west longitude to 53° west); France (Adélie Coast); Norway (Queen Maud Land); New Zealand (160° east longitude to 150° east); and the United Kingdom (British Antarctic Territory).

For several years scientists have been studying the hole in the ozone shield above Antarctica. Without the thin layer of deep blue gas in the upper atmosphere, the sun's scathing ultraviolet rays would wreak environmental havoc on Earth. An equally sharp decline in ozone over New York City, for example, would make sunbathing a death-defying act. Environmentalists have argued that the hole in the ozone is a direct result of atmospheric pollution by chlorofluorocarbons (CFCs) used, for example, in the manufacture of air conditioners, microchips, and fast-food containers. Now, however, some scientists believe

Cold Facts

• Antarctica's mountains are an extension of South America's Andes and complete the broken chain of mountains that begin with the Rockies in northern Canada, become Mexico's Sierra Madre, and continue through Central America to meet the Andes in Colombia. If the ice melted, the mountains would actually be islands.

• Approximately 90 percent of the world's ice and snow can be found in Antarctica.

• Much of Greater Antarctica east of the Greenwich meridian (0° longitude) is actually a high continental shelf, a shallow submarine plain of a continent that typically ends in a steep slope to the ocean abyss. The Antarctic ice cap actually extends the continent's land above sea level.

• The most desolate place on Earth, Antarctica is the only continent that has no flowering plants, no grasses, no large mammals, and no permanent population.

• The greatest recorded depth of the Antarctic, or Southern, Ocean: 21,043 feet, at 66° 58' south latitude (near the Antarctic Circle), 176° 14' west longitude.

that natural chlorine gas may also be contributing to the depletion of the ozone. Mt. Erebus, on Ross Island, last erupted in 1989. It spewed large quantities of chlorine gas, which many scientists believe also contributed to the so-called hole in the ozone layer now present over Antarctica.

In 1991 the entire world community agreed to sweeping protection of one of the last pristine environments on Earth. The Environment Protection Protocol added to the treaty places Antarctica off-limits to development, strictly regulates waste disposal and marine pollution, places restriction on tourism and research, and bans military testing and the mining of fossil fuels.

The Arctic

Unlike the geographical South Pole on Antarctic land, the North Pole lies beneath the Arctic Ocean's year-round seven- to ten-foot-thick ice, where nobody lives, apart from the few transient Americans, Canadians, and Russians during their stints in floating research stations. For centuries polar peoples have made their homes along the Arctic Circle, having learned to cope with the frozen lands and seas in the northern extremes of Siberia, Scandinavia, Alaska, and Canada, as well as two-thirds of Greenland. The North American and Greenland Inuits (Eskimos), for example, are hunters. Others, like the Scandinavian Lapps and the Siberian Chukchis, keep herds of reindeer, which provide meat, milk, fur, leather, and fuel (in the form of dried manure).

Early Arctic explorers sought a route around North America that would nearly halve the sea distance from Europe to the Far East. In 1903-1906, Roald Amundsen, the Norwegian who was the first to reach to the South Pole, made the first successful voyage around the northern coast of North America. But the tanker *Manhattan*, in 1969, was the first commercial vessel to forge directly through the Arctic ice.

World Water

Source	Approx. % of total
Salt water	
Oceans	96.63
Inland seas and salt lakes	0.008
Fresh water	
Antarctic ice cap	1.9
Groundwater	0.63
Within half a mile of surface	0.31
Deep-lying	0.31
Arctic ice cap and glaciers	0.21
Freshwater lakes	0.009
Atmosphere	0.001
All rivers	0.0001

Geofacts

· The greatest recorded depth of the Arctic Ocean: 17,880 feet in the North Polar Basin.

· Less than 5.5 million square miles, the Arctic Ocean is the world's smallest ocean. It has a lower salt content than the other oceans.

· The Lambert Glacier is the largest glacier in the world.

Quiz

(answers, p. 155)

1. What are the earth's magnetic poles and where are they located?

2. Learning to cope with the harsh Arctic climate, the Inuits (or Eskimos) are noted for their unique shelters. What are these shelters called?

3. What is the American submarine *Nautilus*'s claim to fame?

4. What two countries lie closest to Antarctica?

5. Where is the only place you can sail a boat in a straight line—east to west—without hitting land?

6. In which hemisphere can you find penguins? Polar bears?

7. What did Robert E. Peary and Matthew Henson discover in 1909?

8. If you went to Antarctica on June 21, what season would you encounter?

9. How did Manchurian ponies end up in Antarctica?

10. How many independent nations lie at least partly within the Arctic Circle?

11. Where is the world's largest supply of fresh water located?

ANSWERS

1. A compass needle points to the northern or southern magnetic pole from any direction except in the immediate vicinity, where the intensity of the attraction is so slight that a compass can't determine direction. In the Northern Hemisphere, a compass will point to the North Magnetic Pole, while in the Southern Hemisphere, it will point to the South Magnetic Pole. Unlike the geographical poles, the magnetic poles are not fixed in place but move very slowly. In 1987, for example, the South Magnetic Pole lay just off the coast of Adélie Coast, the French outpost in Antarctica.

2. Igloos

3. The *Nautilus* was the first vessel to reach the North Pole by traveling beneath the Arctic ice.

4. Argentina and Chile, which share Tierra del Fuego on the northern side of Drake Passage, are tied for closest.

5. The Antarctic Ocean

6. Polar bears are native to the Arctic. Penguins are found only in the Southern Hemisphere. A variety of penguin species waddle ashore in Antarctica during the summer months, but all of them except the emperor penguin leave before the brutal onset of winter.

7. U.S. Admiral Robert E. Peary and Matthew Henson discovered the North Pole on April 6, 1909. Until recently, history books gave the complete credit to Peary, and mention of Henson, a black man, was relegated to his being a member of Peary's team. Peary, however, considered Henson an equal partner in the venture, and in 1991 Henson's body was moved from a modest grave in Connecticut and laid to rest beside Peary in Arlington National Cemetery, in Arlington, Virginia, outside Washington, D.C.

8. Winter. The seasons in the Southern Hemisphere are the exact opposite of those in the Northern Hemisphere.

9. Pioneering the way to the South Pole, British explorer Ernest Shackleton used Manchurian ponies to pull the sleds during his 1908 Antarctic trek. Unable to withstand the bitter weather, however, the ponies soon died.

10. Six nations do—Finland, Sweden, Norway, Russia (Siberia), the United States (Alaska), and Canada (Northwest Territories). A Danish territory, Greenland, is not an independent nation.

11. Antarctica, which contains nearly 2 percent of all the world's water in the form of ice